The Elder Scrolls V

SKYRIM

ANNIVERSARY EDITION

THE COMPLETE AND LATEST GUIDE

Anders Lindstrom

TIPS AND TRICKS

HOW TO SELL YOUR STOLEN GOODS

Crafting

This one is restricted to a few objects, but is worth considering if you've been lifting from hunters, mines, or blacksmith shops. Items crafted from stolen materials will become legitimate. This includes leather, which can be made into fully legal leather strips, or hides that will become legal leather.

Stolen ore can be smelted into ingots and stolen ingots can be turned into weapons, armour or building materials to avoid them being traced back to a crime.

Tonilia In The Ragged Flagon

Tonilia is the first Thieves' Guild fence that you're likely to encounter. Fences are particular merchants who will let you offload goods that are earmarked as stolen. She can be found in the Ragged Flagon, the gloomy tavern at the end of Riften's Ratway. She doesn't just offer her services to anyone, however. You will need to be a member of the Guild to do business with her, or any of the other fences.

Once you're in with the Thieves' Guild, she will have 1000 septims to buy with, an amount that can increase up to 4000 once the Guild has achieved appropriate amounts of influence in various holds.

Mallus Maccius In Honningbrew Meadery

Mallus Maccius will become the easiest way to offload stolen goods in the Whiterun area. First, you have to get him out of his job as owner Sabjorn's put-upon assistant and into the management of the Honningbrew Meadery.

Placing Mallus in authority is done through the 'Dampened Spirits' quest. On instruction from Maven Black-Briar and with assistance from Mallus, you break into the meadery and do a bit of sabotage. The result will be a mildly-poisoned Guard Captain, Sabjorn in prison, Mallus running the place, Maven off your back, and you being able to finally offload all the gold necklaces in your pockets.

Niranye At Windhelm Market

Niranye is an elf who trades out of Windhelm Market. She once acted as a fence for the Thieves' Guild, and she will again, but there is a group standing in the way. A rival group of criminals, The Summerset Shadows have blackmailed her into working for them.

A quest of the same name as the group, 'Summerset Shadows', will become available when the Guild has enough influence in Windhelm from smaller jobs. In the quest, you'll be charged with questioning Niranye and eliminating the rival

6

group. After that, the Altmer merchant will happily take your stolen goods.

Gulum-Ei In The Winking Skeever

Gulum-Ei is a shifty Argonian, normally found drinking in Solitude's tavern, The Winking Skeever. As part of the Thieves' Guild main quest 'Scoundrel's Folly', you'll be charged with finding out who he is doing business with. He will depart his spot in the Inn and you will need to trail him down to the East Empire Company and through to the back. Following the lizard man into the cave will reveal all sorts of shady deals he's been up to while working at the warehouse, and he'll be a lot more cooperative once tracked down. Spare him and let him work his debt off.

Gulum-Ei is not only a useful fence for Solitude, but he is placed close enough to the Temple Of The Divines that you can pick up the blessing of Zenithar and get better prices easily. Handy!

Enthir In The College Of Winterhold

Enthir is a woof elf who works at the College Of Winterhold, and is known for his ability to acquire all kinds of hard-to-find items. There is a reason he's so good at it the man is on good terms with certain people in the Thieves' Guild. Unfortunately, the people he's on good terms with won't show their faces until you're a little way into the Guild quest line.

Once 'Hard Answers' is completed, Enthir will offer his help as a buyer, and you will be able to trade with him if you're in the area. Very helpful if you're a mage that moonlights as a robber.

Endon In Markarth

You might recognise Endon's family from the rather alarming introduction you get to Markarth. His wife Kerah runs the silver stall just inside the gates, where the stabbing attempt happens. Aside from that, Endon is probably less shifty than most people you're likely to do business with. He comes to the Thieves' Guild when they have some pull in Markarth. He's a bit desperate. Bandits have taken an irreplaceable silver mold from him, but no legitimate authorities will help.

On taking the 'Silver Lining 'quest, you can sneak or fight their way into Pinewatch and find the mold, something that will delight Endon. In turn, he will happily accept stolen goods in return. Who said the Thieves' Guild don't help the community?

Khajiit Caravans

There are three rotating Khajiit Caravans in Skyrim. They move from city to city, but are not permitted to pitch up within the walls. Why? The people of Skyrim have some rather unpleasant preconceived notions about the cat people. They think they're all thieves, and that they traffic illegal goods. Perhaps if they were treated with less suspicion, then it wouldn't be as easy to talk them around into dealing in stolen goods.

When the Thieves' Guild has become sufficiently well known, Tonilia will ask you to deliver some rather illegal moon sugar to caravan owner Ri'Saad. This will be enough to make a deal, and he will spread word to the other caravans to take illegitimate goods off your hands.

Any Shop You Want (With A Perk)

What is better than any trick or fence? Being able to trade any good with whomever you want. How is that done? With the 'Fence' perk. Taking the perk allows you to trade with any merchant as if they were in the Thieves' Guild's pocket, and you don't even need to be part of the Guild to make use of it.

The downside of the perk is that its requirements are pretty hefty. With a 90 in speech needed, and a fair amount of investment into the tree, you will already have had to prove yourself a smooth talker with a lot of Skyrim beforehand.

THE BEST PLACES TO FIND BUILDING MATERIALS

Stone And Clay - Your Own Property

Quarried Stone and Clay are the backbones of your construction efforts. You will need to gather a lot of these if you want to get anywhere in building your Nordic dream home. Fortunately, these are some of the easiest materials to come by. Quarried Stone is needed for floors, foundations, and most major house features, while Clay is required for smelters, plinths, and walls.

Any land you purchase will have both a Stone quarry and a Clay deposit somewhere on the property, and they will output a large number of units before running low. If you don't want to sit and watch your Dovahkiin hit something with a pickaxe for ten minutes, there are other options. If you've appointed a steward to your property, they can purchase Stone or Clay for you, at 20 Septims for 20 blocks of Clay, or 100 Septims for 20 chunks of Stone.

Sawn Logs - Lumber Mills

One of the other major cornerstones of your building projects is Sawn Logs. These can be bought from one of the various lumber mills around Skyrim, at 200 Septims for 20 Sawn Logs. If you've appointed a steward, they can do the legwork for you, then you can buy the logs from home.

If money is an issue, it's possible to operate the mill yourself and saw your own logs with each cut giving you ten for your log pile. You can't just jump in though, as you'll need to befriend the lumber mill owner first. This can be done by cutting a lot of firewood for them, though some also have quests that will help. Aeri of Anga's Mill needs a letter delivered to Morthal, and Horgeir at Dragon Bridge will appreciate it if you return his Dragon's Breath Mead to him.

Straw - General Goods Stores

Straw is a useful commodity that comes up in a few plans for furniture, including beds and archery targets. It's also not necessarily obvious where you can get hold of it.

Straw will show up in general goods stores, the ones with scales on the signage. Belethor's General Goods in Whiterun is one of these, though it's advisable to look in a few of them as there's no guarantee of Straw being in stock on any given day. It can sometimes be stolen from houses and stores as well.

Iron Ingots - Mines and Smiths

Iron Ingots are needed for nails, hinges, and most other metal pieces used in construction. They can be a bit of a pain to come across in large quantities.

Blacksmiths may sell a few at a time, or you can stop off at an Iron mine. Left Hand Mine near Markarth and Stonehills in Hjaalmarch are both convenient, as they have Iron and a smelter right outside. If that all sounds like too much hard work, and you don't mind less-than-legal ways to get your hands on it, The Scorched Anvil in Riften has a big stack of ingots on the lower floor that you could liberate. They probably won't miss them too much.

Corundum - Darkwater Crossing

Corundum is more valuable and noticeably harder to find in ingot form than Iron. Fortunately, it's also not needed in anywhere near the same quantities. Most of what you will need it for is making locks, which require one ingot and are often needed when adding additional rooms. A good source is Darkwater Crossing, as it has a smelter on-site and you're unlikely to be bothered while you're there.

If you're the larcenous kind of player, you can also pick up a few ingots that sit out in the open by the smelter — especially if the miners have gone to bed. Well, they shouldn't leave them lying around if they don't want someone to pick them up.

Glass - Shops And Caravans

Despite the name, Glass in Skyrim is not used to make glass weapons — they're made from Malachite, which does make you wonder about the name. Was it too long for the Nords?

Items in your house, on the other hand, are made from actual Glass. It can be found in cupboards in homes, whether taken legally or lifted while the owner isn't looking. It will sometimes pop up in general goods stores' inventories, and can also often be sold by the Khajiit caravans — if you can find one camped out.

Pelts - Hunting Animals

Pelts of some kind are used to craft mounted animal heads. The important thing to keep in mind here is that you will need the respective pelt of the animal you want a trophy of. As many animal encounters in the wild depend on your level, things like bears and sabrecats might be a little trickier to find early on. You can have a go at taking on the Helgen cave bear if you want a trophy to show off as early as possible.

If you're less worried about species specifically, The Rift is a good place to hunt, as its birch forests are invariably teeming with wildlife. Finally, a reason to actually want to be mobbed by the super-aggressive wolves of Skyrim.

Leather - Blacksmiths

Leather Strips are used in a number of household items, as well as being helpful in some weapon crafting. They're made from Leather, which is in turn made by working Hides on a tanning rack. If you have plenty of skins stocked up, it might not be an issue, as there's a tanning rack available at your property fairly early on.

If you don't have Hides on hand, it's probably easier to head over to the nearest blacksmith. They always have Leather Strips in stock, and they're very cheap, so there's no reason to take a whole hunting trip unless you want to.

Goat Horns - Goats

No prizes for guessing where you get Goat Horns from. If you want to make lighting fixtures, a goat is going to have to go and meet the Divines. You can find these passive herbivores wandering wild around Skyrim, and you will almost inevitably find at least one on a farm.

Keep in mind that goat murder is taken quite seriously in the northern kingdom, so being seen attacking one will earn you a bounty in the local hold. If hunting isn't fruitful, and you don't want to embark on a life of crime, you can sometimes find them for sale at general goods stores.

SHADOWMARKS AND THEIR MEANINGS

The Guild

This mark indicates a safe house, entrance, or contact with the Thieves' Guild and indicates that the area is friendly to or serves as property for the Guild. One of the most notable examples is the sarcophagus sign that acts as the secret back entrance to the Ragged Flagon's Cistern.

It can also be seen on the doorway to Riftweald Manor, which is owned by the sometime Guildmaster, Mercer Frey. If you're looking to sign up with them, this sign is a good indication you're on the right track.

Safe

The Safe symbol indicates routes that are safe to use. They can either indicate good areas to stay unseen or point toward safehouses associated with some members of the Guild. They're one of the less-commonly seen marks, though notably, one instance does foreshadow an important aspect of the Thieves Guild plot.

You can find one to the south of Riften, which points to the entrance of Nightingale Hall, though oddly, you won't be able to reach it until accessing the corresponding quest in the Thieves Guild plotline. At least it'll make it easier to find when the time comes.

Danger

The 'danger' mark indicates theft locations that the Guild really doesn't consider worth the risk, and it often denotes very strong fighters in residence. This is the reason that you can find one by the Jorrvaskr entrance or Aerin's house in Riften, where Mjoll the Lioness resides.

The mark can also relate to different kinds of threats. There is one on the Arentino residence in Windhelm, presumably as Aventus is trying to perform the Black Sacrament inside. Even members of the Guild don't really want the Dark Brotherhood on their tail.

Protected

This mark indicates anyone in an arrangement with the Guild. Individuals, businesses, or families whose properties have these marks outside should not be robbed or harmed. They can be seen, unsurprisingly, on Black-Briar Manor, though also interestingly on Erikur's house in Solitude and the house of Clan Battle-born in Whiterun.

Despite the threat attached to this mark, there don't seem to be in-game consequences for violating the terms of the mark. There's only the ever-present threat of getting attacked by hired thugs sent by the former owner of the items. Maybe the other members are spitting into your beer in the Ragged Flagon.

Escape Route

One of the most useful Shadowmarks if you're the kind of player who gets into trouble frequently or can't resist launching Nazeem right over Dragonreach with a strike from Volendrung. You will often find these inside jail cells, and they are indicators of a way to make your grand escape from confinement.

Escape route marks will also show up in a few other places. You can use one to make a swift exit from the back of Riftweald Manor into the Ratway, presumably added by Mercer as a way to get to the office in a hurry.

Loot

One of the more common Shadowmarks you're likely to find, these will be left by any property that the Thieves Guild considers to be worth stealing from. This means there are sufficient valuables to make it worth it, and it's not hazardous enough to warrant a 'danger' mark.

Places with the 'loot' mark include many of the shops in Skyrim, as well as houses of wealthy individuals who clearly haven't managed to pay the Guild off in the manner of the Black-Briars. Seeing this is a good sign that you should go in and take a look at what's on offer.

Empty

'Empty' is a fairly self-explanatory mark, and it denotes an area or property that doesn't have anything worth taking. This is usually found outside the properties of people who are down on their luck. Places like Beggar's Row in Riften and The Warrens in Markarth have these marks.

While they're generally a reliable indicator that you won't find anything expensive inside, occasionally, this mark can be misleading. The 'empty' sign on the abandoned house off of Markarth's square is far from accurate, though maybe it was warning off people from something rather nasty inside.

Fence

A useful sign to remember if you make a habit of taking what isn't yours, a fence sign, unsurprisingly, denotes the presence of a guild fence in the area. These are fine individuals who will purchase stolen goods from you, whereas other shopkeepers won't touch such a thing (without a particular perk in effect).

Fences are only available through the Guild, and initially, the only fence will be Tonilia in the Ragged Flagon. Completing quests will broaden the number available to you, and eventually, there will be convenient locations in various cities to offload ill-gotten gains.

Thieves' Cache

You never get something for nothing, right? Wrong. All you need to do is keep an eye out for this symbol. It will often appear on containers in cities that the Thieves' Guild has been actively working on.

Opening caches up will reveal some surprisingly valuable leveled loot, including enchanted weapons, gold, lockpicks, and jewels. Even better, it's all free to take, and no one will try and stop you, and it's not considered stolen goods. Just a little present from your fellow rogues. Who said there's no honor among thieves?

THE SHRINES OF SKYRIM

Stendarr - Become A Mighty Adversary

God	Stendarr	God Of...	Righteous Might Merciful Forbearance
DLC Required	None		
Effects Upon Activation	Cures all diseases (except advanced Vampirism and Lycanthropy) For eight hours, you will block ten percent more damage		

As the God of Mercy and Justice, Stendarr is a Divine that has a loyal following of vampire hunters. The Shrine of Stendarr, which typically resembles a small horn pouring liquid, will give a bonus to your Block skill.

If you run a playstyle where Block is important, this will come in handy. Otherwise, it's of basically no use, making it one of the worst shrines in the game.

Mara - Embrace Your Inner Romantic

God	Mara	God Of...	Love
DLC Required	None		
Effects Upon Activation	Cures all diseases (except advanced Vampirism and Lycanthropy) For eight hours, Restoration spells use ten percent less mana when you cast them		

Mara is a divine that is called the Mother Goddess. She rules over the power of love and plays an important role in the marriage process in Skyrim.

Her shrine, which has a large cross with a face in the center, will grant the player a ten percent bonus to healing spells. Again, this is useful for people who specifically focus on Restoration. Otherwise, it doesn't have a ton of universal use.

Julianos - You Have To Be The Nerd You Want To See In The World

God	Julianos	God Of...	Wisdom/ Logic
DLC Required	None		
Effects Upon Activation	Cures all diseases (except advanced Vampirism and Lycanthropy) For eight hours, your Magicka is increased by 25 points		

Wisdom and logic are the domain of Julianos. Some of his followers are in charge of keeping the Elder Scrolls themselves, due to Julianos' nature as the god of intelligent endeavors.

His shrine, which looks like an intricately-designed pyramid, will increase your Magicka by 25 points. For mages and other Magicka-using builds, this is essential. Other than that, it won't do much. This bonus is also very common on rings and armor.

Boethiah - Cue The Ominous Music

God	Boethiah	God Of...	Deceit Conspiracy Treachery Sedition
DLC Required	Dragonborn		
Effects Upon Activation	Cures all diseases (except advanced Vampirism and Lycanthropy) For eight hours, your One-Handed weapons will do ten percent more damage		

Boethiah is a Daedric Prince, not one of the Nine Divines. This deity rules over murder, assassinations, treason, overthrowing authority, deceptive acts, and conspiracies. His shrine can be found next to the shrines to other Daedric Princes Mephala and Azura in the Raven Rock Temple.

When used, it'll make One-Handed Weapons do ten percent more damage. If you dual-wield One-Handed weapons, this bonus can be deadly.

Nocturnal - Lady Luck Is On Your Side

God	Nocturnal	God Of...	Night Darkness Luck
DLC Required	None		
Effects Upon Activation	Cures all diseases (except advanced Vampirism and Lycanthropy) For eight hours, you will be ten percent harder to detect when you are crouched and sneaking		

Another Daedric Prince, Nocturnal rules over the night, darkness, stealth. Her shrine can be found in the Thieves Guild's hideout under Riften, which makes sense considering she's also known as Lady Luck.

Nocturnal's shrine will, unsurprisingly, give a bonus to Sneak. This bonus is particularly helpful to stealthy builds, but it is useable by almost anyone, other than maybe someone who loves to run into things guns a-blazin'.

Auriel - For The People Who Have Everything

God	Auriel	God Of...	Everything Eternal Immortal Heaven
DLC Required	Dawnguard		
Effects Upon Activation	Cures all diseases (except advanced Vampirism and Lycanthropy) For eight hours, your attacks with missile weapons (bows and crossbows) do ten percent more damage		

Akatosh, one of the Nine Divines, has an Elvish counterpart, Auriel. This Elvish deity is considered the leader of the pantheon, and is seen as part of "The Everything."

Auriel's shrine, which resembles a large sun, is only found in two places in all of Skyrim: Darkfall Cave and the Inner Sanctum, and only when the player has the Dawnguard DLC. When used, it will confer a blessing that gives a ten percent boost to bows and crossbows.

Akatosh - Live Up To The 'Dragonborn' Title

God	Akatosh	God Of...	Endurance Invincibility Everlasting Defender
DLC Required	None		
Effects Upon Activation	Cures all diseases (except advanced Vampirism and Lycanthropy) For eight hours, your Magicka will recharge ten percent faster than normal		

Akatosh is Auriel's variant in the Nine Divines, according to the Imperial Cult. Like Auriel, he is seen as the leader of his respective pantheon. He represents endurance, invincibility, everlasting legitimacy, while also ruling over time, often manifesting as a dragon.

His shrine will bless the player with Magicka Regeneration. Specifically, Akatosh's blessing will make Magicka regenerate ten percent faster. This will give a bonus to any mage of any school, instead of the school-specific blessings of other shrines.

Kynareth - Skyrim: The Last Airbender

God	Kynareth	God Of...	Winds Elements Air Heavens Sky

DLC Required		None
Effects Activation	Upon	Cures all diseases (except advanced Vampirism and Lycanthropy)
		For eight hours, your Stamina is increased by 25 points

Kynareth is the goddess of nature and natural elements, as well as some aspects of the sky and air. Her shrine looks like a bird, owl, or wings on a pedestal.

Using this shrine will give the Blessing of Kynareth, which grants an extra 25 Stamina points to the player. This is useful for those who use Stamina-heavy builds like a Two-Handed warrior. But, it also has a use for almost anyone, as Stamina is consumed when running, attacking in some ways, and bashing.

Dibella - Talk Anyone Into Or Out Of Their Pants, Your Call

God		Dibella	God Of...	Love
				Beauty
				Music
				Art
DLC Required		None		
Effects Activation	Upon	Cures all diseases (except advanced Vampirism and Lycanthropy)		
		For eight hours, your Speech skill gets increased by ten points		

Dibella is the Imperial Cult's goddess of beauty, love, and artistry. Her shrine looks like a lotus flower with petals opening. They can be found in a variety of places, including Riften and Solitude.

This shrine will grant the Blessing of Dibella, which gives an extra ten points to Speech. This will allow you to have better haggling skills with merchants, persuade people more easily, and more. This has use for any type of build in the game.

Azura - When You Have To Get Up At Six To Catch The Bus

God			Azura	God Of...	Dusk and Dawn Transition Change
DLC Required			Dragonborn		
Effects Upon Activation			Cures all diseases (except advanced Vampirism and Lycanthropy) For eight hours, your Resist Magic skill is increased by ten points		

Alongside the Shrine of Boethiah in the Raven Rock Temple is the Shrine of Azura. This Daedric Prince is the deity that rules over dusk, dawn, and the magic realms they inhabit in between the twilight.

Azura's Blessing will give you a ten percent higher resistance to Magic attacks. This is useful for almost anyone. No matter your player build, you'll have to contend with Magic users across the province of Skyrim.

Mephala - A God Who Can Get You A Deal

God		Mephala	God Of...	Lies Sex Murder Secrets
DLC Required		None		
Effects Upon Activation		Cures all diseases (except advanced Vampirism and Lycanthropy) For eight hours, all items cost ten percent less		

The final Daedric Prince on this list, Mephala, is the ruler over lies, sex, secrets, and murder (although many believe this is wrong and that her sphere isn't visible to mortals). She likes to interfere in the matters of mortals for amusement.

It makes sense, then, that her Blessing will give make prices ten percent better. Sure, this might be only part of what the Blessing of Dibella offers, but Mephala is just way cooler, so she gets bonus points.

Zenithar - A Working Man's God

God	Zenithar	God Of...	Work Commerce Trade
DLC Required	None		
Effects Upon Activation	Cures all diseases (except advanced Vampirism and Lycanthropy) For eight hours, all items cost ten percent less		

As the god of work and commerce, it makes sense that Zenithar's shrine looks just like a blacksmith's anvil. He believes his followers should never steal and that they should be smart with their money.

As a way of helping with this, maybe, his blessing will make all prices ten percent better. This is the same as Mephala's Blessing, of course, but Zenithar's shrine places higher on this list simply because it is more ubiquitous. Mephala's shrine can only be found in one place, while Zenithar's can be found in numerous places and even added to your house.

Arkay - Replacement For Health Potions

God	Arkay	God Of...	Birth Death
Effects Upon Activation	Cures all diseases (except advanced Vampirism and Lycanthropy) For eight hours, your Health is increased by 25 points		

The god of death and the cycle of birth, Arkay represents life in many respects. As a result, his blessing will improve your life, giving you an extra 25 Health points when you activate a Shrine of Arkay.

This is obviously very useful to any player in any type of playstyle. This is actually a very useful shrine to command a follower to use, as well.

Talos - Yell Louder And More Often

God	Talos	God Of...	Just Rule
			Civil Society
Effects Upon Activation	Cures all diseases (except advanced Vampirism and Lycanthropy)		
	For eight hours, the length of time you need to wait between Shouting is 20 % shorter		

Regardless of playstyle or character build, the lore of Skyrim dictates that the Dragonborn, the player character, is the wielder of the power of Shouts. These Shouts have a wide variety of uses, but one thing remains true: they have to recharge.

Whether you're trying to use a combat Shout to best your foes or just using Unrelenting Force to blast your follower off a cliff for fun, waiting for Shouts to recharge is tedious and annoying. The Shrine of Talos, the god of good governance and just authority, will help with this, reducing the time between Shouts by 20 percent. Regardless of his outlaw status, Talos shrines can be found around Skyrim.

THE BEST MINES TO DIVE INTO

Embershard Mine

One of the earliest mines you're likely to run into on your Skyrim journey is the Embershard Mine. It sits just off the road between Helgen and Riverwood and is a great place to pick up some extra cash before heading to Whiterun.

Bandits have taken over the place, and a few lurk inside, though they are positioned in places that make most easy enough to sneak up on. A little way in is a large chest in a locked room, along with the spell tome for Clairvoyance, which can be grabbed without getting through the gate.

Bilegulch Mine

Right at the border of The Reach and Falkreath holds, you can find Bilegulch Mine. Your approach is not likely to warrant a warm welcome. The mine, and the surrounding fort, have been taken over by orc bandits. Ploughing through these hostile individuals ends with the Chief, who is holed up in the mine itself. Once clear, the place becomes a great place for smithing orc weapons.

The mine has numerous orichalcum veins, and the fort at the top has a full set of smithing tools that will be free to utilize with the bandits taken care of.

Sanuarach Mine

On walking into the village of Karthwasten, you can see a disagreement taking place. This is between Ainethach, a Reach native who owns the local Sanuarach Mine, and a group of mercenaries hired by the Silver-Bloods. The mercenaries refuse to let anyone mine, supposedly as some means to protect the area from Forsworn. Ainethach is pretty sure it is simply the Silver-Bloods trying to strong-arm him into selling up.

You can go in and convince the mercenaries to pack up and leave, or you can get rid of them in a much less subtle way. Alternatively, you can persuade the owner to sell up. Whichever way you go about it, you can pick up a decent gold reward for your troubles.

Lost Prospect Mine

Not far from the Black-Briar Lodge, you can find Lost Prospect Mine, a small shaft seemingly deserted of activity. There you can find a journal from a man named Hadrir, lamenting the failure of the mining venture. He also notes that his associate Bern has apparently left.

There are a few short shafts and a waterfall. The use of Whirlwind Sprint, or some creative jumping, can allow the Dragonborn onto the top of the waterfall. Following the stream back will lead to a small passage, and eventually, what appears to be the skeleton of Bern, surrounded by gold ore veins. It appears he found exactly what they were looking for.

Raven Rock Mine

A little off the main square of Solstheim's capital Raven Rock, you can find the abandoned ebony mine. On the first entry, you're likely to hear two people having a disagreement there. This is Crescius Caerellius and his wife Aphia. Crescius wants to descend into the mine. His grandfather died there, and he is convinced that the East Empire Company covered up the reason. Going down in his stead will soon reveal he was completely right.

The shaft breaks into an ancient barrow, where the strange, unique weapon, The Bloodskal Blade, can be discovered. This sword releases a wave of energy through the air on a charged attack.

Kolskeggr Mine

This place is best approached after paying a visit to the nearby Left Hand Mine. There you will find Pavo Attius and Gat gro-Shargakh, the only survivors after a band of Forsworn attacked Kolskeggr Mine.

Speaking to Pavo will lead him to ask you to expel the intruders in the most decisive way possible. In the mine itself, you can find a band of Forsworn, led by a Briarheart. This can sometimes be a difficult fight, but it is possible to sneak around and take out the Forsworn individually to avoid being swamped. Pavo will be delighted by the outcome and give you a reward.

Soljund's Sinkhole

A short walk from the Old Hroldan Inn is Soljund's Sinkhole, a mine with a bit of a draugr problem. According to Perth, one of the miners accidentally dug into an old Nordic crypt, and the tunnels became flooded with angry draugr. Fortunately for them, you can offer to clear out their mine.

After getting through the initial shafts, it is possible to find the drop into the tomb itself. On entering, there will be a few traps, and at the end, a powerful draugr, along with a statue that will pelt fire at you. Once the boss draugr and all his minions are taken care of, you can return to Perth for some gold. Don't forget the chest in the tomb, too.

Gloombound Mine

One of the few ebony mines in Skyrim proper, Gloombound mine is attached to the orc stronghold of Narzulbur. If you are an orc or have aided an orc sufficiently to get the status of blood-kin, then you will be able to enter and wander around freely. If you are neither of these, you may have to sneak in to avoid upsetting the locals.

As an ebony mine, there are plenty of veins of the valuable mineral scattered around, waiting to be mined. It is not without its dangers, however. Some floors will collapse into tunnels below. More pressingly, there are areas where the air seems to wobble. This is a flammable gas and may ignite if you use a torch of fire magic. Take care.

Northwind Mine

Northwind Mine can be found on the edge of The Rift, near Shor's Stone. It is a small mine abandoned by miners, though not wholly unoccupied. The place is full of wandering skeletons who will attack on sight but are not too difficult to deal with

Making your way through will lead up to Northwind Summit, where you can find more skeletons, a big chest, and a Word Wall for Aura Whisper. If you don't feel like taking on the mine part, it's possible to drop down on the summit from the world map with a bit of careful jumping.

Cidhna Mine

Of all the mines in Skyrim, Cidhna Mine is a little unusual because you're not likely to visit willingly. Run by the Silverblood family, it functions as both Markarth's silver mine and its prison. Convicts are thrown into the bottom and expected to provide ore in exchange for food. If you get too closely involved in some of the city's shady dealings, you may be thrown down there and have all your belongings taken away.

Escape from the mine by either befriending or killing Forsworn King Madenach, also imprisoned inside. Either option will reward you with unique items, an enchanted Forsworn set from Madenach or a barter-increasing ring from the Silver-Bloods for dealing with their problem.

UNLOCK EVERY HEARTHFIRE HOME PET

What Are Pets?

Unlike animal companions in Skyrim, pets don't offer any major advantages. You'll need to adopt children before you can recruit a pet. However, they make your children happier and will walk around your house, adding a unique ambiance to the building.

You can only have two pets in total, so choose carefully. You can recruit pets in a variety of ways, all of which are detailed below.

Random Dog

You'll occasionally encounter a random dog throughout Skyrim that you can recruit as an animal companion. Once it's following you, bring the dog to a Hearthfire home with children, and they'll ask to adopt it, which doesn't require any additional effort like feeding the pet. This encounter is random, and it's possible to find stray dogs in Skyrim and on Solstheim.

Meeko

Meeko is a unique dog that you can recruit for free to act as your animal companion in combat. However, you can also bring it to your Hearthfire home if you have children and adopt it for them as a gift. It will then act as a permanent pet for your family.

Recruiting Meeko is simple because all you have to do is travel to Meeko's Shack near Solitude, which is on the map above. It's also possible to first encounter Meeko on the road nearby the shack. Once you encounter it, you can ask it to follow you for free.

Vigilance

Vigilance looks exactly like Meeko, but instead of encountering it in the wild, you'll need to purchase this dog for 500 Gold at the Markarth Stables, just outside the main gate.

After purchasing Vigilance, it'll follow you like any other animal companion and you can bring it to one of your children at a Hearthfire home to adopt it. Vigilance requires no extra upkeep after agreeing to keep it as a pet.

Bran And Sceolang

Dawnguard Huskies are a rare breed of dog that you can recruit as animal companions if you side with the Dawnguard during the "Bloodline" quest. You'll next need to complete the "A New Order" mission, which will then allow you to command Bran and Sceolang to follow you.

After recruiting Bran or Sceolang as a follower, travel to your Hearthfire home and interact with your children. They'll ask to keep one of the huskies and the dog will then walk the grounds of your home permanently without any upkeep necessary.

Garmr And CuSith

Garmr and CuSith are two Death Hounds living inside Castle Volkihar in the Dawnguard DLC. Both of these dogs are adoptable as pets with the Hearthfire DLC but you'll need to first complete "The Bloodstone Chalice" quest, which is the first mission you'll receive from Lord Harkon after joining the Volkihar Vampires during the "Bloodline" quest.

Although Garmr and CuSith have very aggressive appearances, your children don't mind and will still want to adopt them as pets should you command one of them to follow you to your Hearthfire home. Upon adoption, they'll simply walk around your house without needing to be fed.

Mudcrab

The Mudcrab is named Pincer, and unlike Mudcrabs you can encounter in Skyrim, this pet isn't hostile at all. Only a male child can recruit Pincer, so having one is a requirement for unlocking this pet.

Eventually, after adopting a male child and moving to a Hearthfire home, he may bring home Pincer and ask to keep it permanently. Pincer will follow the child around the house but doesn't require you to feed or look after it.

Snow Fox

Vix is a Snow Fox of the same variety that you'll encounter throughout the snowy regions of Skyrim. No effort is necessary after Kit is your pet, but it will stroll around your home. You'll need to adopt any female child in Skyrim, then wait until she brings Vix home and asks to keep it, after which it'll become your pet permanently.

Red Fox

The Red Fox you can adopt is named Kit and it's one of the smallest pets in the Hearthfire DLC. Like other pets, upon adopting Kit, it doesn't require food or anything else, it'll simply follow the child around your Hearthfire house. You need to adopt a female child and eventually, she may bring home Kit and ask to keep the fox as a permanent pet.

Skeever

Biter is a unique Skeever that's slightly smaller than one you'd encounter in the wild. Male children that you adopt and move to a Hearthfire home will occasionally come home with a pet and ask to keep it, of which Biter is one of the possible pets.

Biter will follow your child around the grounds of your house if you agree, with no additional work necessary to feed or look after it.

Rabbit

Cotton is a Rabbit that any female child you adopt may bring home after playing outside at your Hearthfire home. She'll ask you to keep Cotton, and you can choose whether to adopt the Rabbit or not. If you choose to adopt Cotton, then it'll follow behind your child while she walks around the Hearthfire home, but it won't require any extra work.

MAIN QUESTS

UNBOUND

How To Escape The Dragon

That was wild, huh? That dragon may have accidentally saved your life, but you only have a few seconds before it takes it. Follow Ralof and run to the watchtower ahead. In all the chaos, listen to what they're saying and follow the quest objectives.

This section is quite linear, but you can still die if you're not careful.

Once they finish talking, head up the stairs. Don't worry too much about the snapping dragon head, and keep moving as soon as you can. You'll reach a part of the map that's blocked off by debris. Listen out and make a running jump. You'll miss otherwise. From here, walk through the house and fall through to the first floor. You'll meet Hadvar, the one who asked you for your name.

Follow Hadvar and pay attention to your surroundings. Make sure to hug the wall. The dragon will kill you if you walk into its fire breath. Make sure to follow your former executioner away from the chaos and to the keep. Here, you have to make a decision about who to follow.

- Follow Ralof and join forces with the Stormcloaks
- Follow Hadvar and join forces with the Imperials.

This decision will have some consequences for what you encounter in Helgen Keep, who you talk to in Riverwood and Civil War Questline, but it won't affect the overall game.

How To Get Through Helgen Keep

When you get inside, your chosen companion will unbind your hands, letting you interact. Loot the nearby chests for equipment. They will provide you will armor and weapons to survive the rest of your escape.

After a short delay, you'll be able to leave the room. Follow the path to the next encounter. Here, you will have your first taste of combat against two enemies. Make sure to block incoming attacks and constantly swing at the enemy to chip away at their health. After you defeat them, loot them. They can provide you with equipment to sell or two-handed weapons, if you're interested in going down that route.

Don't be afraid to scavenge equipment from fallen enemies. It's an excellent source of early money.

❖ Collecting Potions

From here, follow your companion out of the room and down the stairs. After a quick interruption from the dragon, your companion will enter a room populated by two enemies. You will get an optional objective to loot a barrel for potions, which you should absolutely do. Make sure to search the shelves; you should be able to find a couple of other potions there.

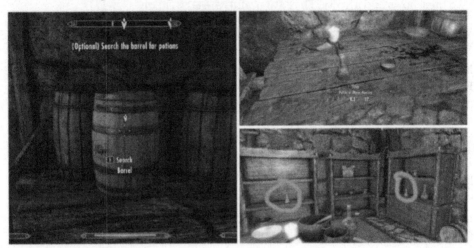

The Torture Room

The next room is the torture chamber, where you'll encounter the Torturer, an Imperial mage, fighting a pair of Stormcloaks. What happens here depends on which person you chose to escape with.

- If you picked Ralof, you will have to fight the Torturer. His shock magic can be dangerous if he chooses to focus on you. Try to hit him with a Power Attack to stun him and prevent him from shocking you.

- If you picked Hadvar, the Torturer will fight by your side, quickly killing the two Stormcloaks without much fuss.

❖ Looting Magical Equipment

A pair of images in a dungeon showing how to get magical loot

Regardless of the outcome, search the knapsack at the center of the room. It will have some lockpicks and a minor potion of healing. Pick the lock on the middle cage at the back of the room to loot a deceased Mage for his robes, septims, and spell tomes. The armory at the back of the room has an Iron Shield and Iron Mace.

The robes and spell tomes are especially useful if you plan on using magic.

Leaving Helgen Keep

The next area is a large, open room filled with enemies. While it's going to be the largest fight that you've had yet, be mindful of the archers at the back of the room. They can mess you up if you're not careful. Odds are your allies will distract nearby enemies, giving you a chance to eliminate the archers. If you're struggling, use the Flames spell to set fire to the oil at their feet, burning them to death. Make sure to take their bows and arrows. They are useful ranged weapons.

From there, you will need to pull a lever to lower a bridge. Once you cross, you won't be able to return.

By now, you'll be seeing signs of nature thanks to the small stream in the cave. Follow it to a cave full of Frostbite Spiders. These enemies will spit venom at you, damaging you over time. They shouldn't be too challenging but if you're struggling to kill them, try shooting them with a bow while sneaking to thin the herd. Keep going once you're done.

The final challenge is a sleeping bear. Your companion will advise you to sneak past it, which is pretty good advice, given how dangerous they are. Still, if you want to slay it, try starting with a sneak attack using your bow before fighting it however you prefer. Those potions may come in handy, as well.

Once you're done, head out and into Skyrim. Congratulations!

BEFORE THE STORM

How To Get To Riverwood

If all you want to do is book it to Riverwood as fast as you can, just follow the quest marker there. Doing so would be a waste of the opportunities you can find along the way, though. As it stands, you're a little on the weak side right now. Without any skills leveled up and holding on to the equipment scrounged up from Helgen Keep, you'll want to get some experience and new gear.

You can really do whatever you want. Join up with the Stormcloak Rebellion or Imperial Legion to take part in the Civil War, visit any of the Hold Capitals, join a faction or just wander. It's all up to you.

The best way to do that is to follow your companion down the path north until you reach a cobbled road. Continuing north until you reach a river will lead you to the Guardian Stones. They are Standing Stones, magic monuments scattered throughout Skyrim that offer unique benefits to the Dragonborn. These Standing Stones offer benefits that are perfect for starting out.

Warrior Stone	Level combat-related skills 20% faster

Thief Stone	Level stealth-related skills 20% faster
Mage Stone	Level magic-related skills 20% faster

The most obvious choice is to pick the stone that matches what your Dragonborn will specialize in. For example, if you want to wear heavy armor and swing two-handed swords, the Warrior Stone would make you excel in that area more quickly.

If, on the other hand, you want to have a Dragonborn that uses skills from across the three fields, such as fighting with one-handed swords mixed with destruction magic, pick the Stone that matches what you'll use least.

At this level, your limited Magicka reserves would make it harder to level your magic-related skills, making the Mage Stone the better choice.

Embershard Mine

Riverwood is waiting just down the road, but we can ignore it in favor of visiting Embershard Mine. You'll be able to see it represented on your compass as a crossed pickaxe and hammer to the east of the Guardian Stones. There will be a hostile bandit waiting for you at the entrance. It won't take long to make quick work of them.

Save before entering and good luck.

Inside, you'll find bandits to loot, chests to plunder, and iron ore if you want to smith your own equipment. If you're stuck at the raised bridge, the solution is to look for a lever found in a room found at the end of the path found on the right of the raised bridge.

Look out for the Clairvoyance spell tome in the locked room after the drop bridge. It's a rare illusion spell that lays out the shortest path to your quest objective.

The rest of the mine should be easy to navigate, and the enemies should not be too challenging as long as you don't try fighting all of them at once. With luck, your skills will have improved from clearing out the dungeon. You should be able to see Riverwood when you head out.

What To Do In Riverwood

The small riverside town may seem like a sleepy backwater, but it has some interesting encounters to offer. Regardless of how you get to Riverwood, you'll need to find your companion's family member.

- If you escaped with Hadvar, you will need to find his uncle, Alvor.

- If you escaped with Ralof, you want to find his sister, Gerdur.

Interact with them to start a conversation where they'll tell you to bring word of the dragon attack to the Jarl of Whiterun. From here, the quest will update, giving you a marker to the city. You may ask them for supplies to help you reach Whiterun. They'll share valuable items, including healing potions, if you want to take advantage of their generosity. You could head there immediately, but you could also explore the town to prepare for the journey ahead.

- Speak to Faendal or Sven about their ongoing feud over Camilla Valerius, starting a short quest.
- Learn basic blacksmithing from Alvor
- Sell your extra equipment to Alvor or Lucan Valerius at the Riverwood Trader.
- Talk to Lucan Valerius about his recent robbery, starting The Golden Claw quest

Starting The Golden Claw will let you skip a large portion of the next Main Story Quest.

With your business in Riverwood concluded, follow the river north to Whiterun.

How To Get To Whiterun

The Hold Capital is hard to miss. Its keep, Dragonsreach, towers over the plains surrounding the city. Follow the river until you reach a crossroads and go west. Along the way, you may meet a group of warriors fighting a giant at a farm. Intervene to earn a little respect from them and an introduction to their guild, The Companions of Jorrvaskr.

Keep following the road to reach the city gates, where a guard will challenge you. Telling him that Riverwood sent you or that a dragon attacked Helgen will be enough to get you in. Once there, head through the city and keep going until you reach Dragonsreach where you can find the Jarl. After some dialogue, he will thank you and give a reward, ending the quest.

BLEAK FALLS BARROW

Prerequisites To Complete

Below, you can find the two prerequisite quests that need to be completed before starting Bleak Falls Barrow.

Quest	How To Start
Unbound	This quest will automatically start after you make your

Quest	How To Start
	character.
Before The Storm	Speak with Hadvar Located in Riverwood.

Accepting The Bleak Falls Barrow Quest

To accept this quest, speak with Balgruuf the Greater, who is the current Jarl of Whiterun; you can find him within Dragonsreach. Upon speaking with the Jarl, he will tell you to meet with Farengar Secret-Fire, the court wizard.

This quest is very closely related to The Golden Claw quest, which requires you to visit Bleak Falls Barrow as well. You can complete these quests at the same time. If you have already finished the Golden Claw, you may have the item needed for the Bleak Falls Barrow quest. In this case, you will just need to talk with Farengar again.

As you speak with Farengar, he will tell you about the Dragonstone, which can be found in an ancient Nordic tomb. He will not provide any additional information, so you will need to venture here yourself.

Traveling To Bleak Falls Barrow

After speaking with Farengar, head to Bleak Falls Barrow, located to the west of Riverwood.

If you have yet to accept The Golden Claw, you can speak with the owner of the Riverwood Trader. Speak with him about a stolen item, and then the quest will begin.

The path to Bleak Falls Barrow is pretty simple; head to the mountains from Riverwood while following the marker on your compass bar. Outside of Bleak Falls Barrow, you will encounter several bandits; defeat them, and then head to the entrance of the tomb.

Inside The Tomb

Entering the tomb will place you in a large chamber containing additional bandits. At this point, if you have yet to accept The Golden Claw, you will receive it after listening to the bandits' conversation.

Defeat or sneak by these bandits, and then continue through the tunnels until you come to a puzzle room with an additional bandit. You will see the bandit trigger a

31

trap by pulling a handle, which shoots darts and kills him. Avoid this trap by solving the puzzle that this room presents. You will need to do this by setting the pillars to the correct symbol. From left to right, the pillars should display the following symbols.

- Snake
- Snake
- Whale

With the pillars set, pull the lever and the door will open. Go through the door and proceed until you find yourself at a spiral staircase leading downwards.

Saving Arvel The Swift

At the bottom of the stairs, you will notice spiderwebs, indicating that the next room will contain a spider. While entering this room, you will also hear the voice of someone in need; kill the spider and then speak with Arvel the Swift, who is stuck in a spiderweb.

Upon speaking to him, he will tell you about the Golden Claw, and promise to help you. Unsurprisingly, Arvel betrays you once you cut him from the webs. You can choose to kill him or let him run; regardless of what you pick, Arvel will die. If you kill Arvel, be sure to loot his body to obtain the Golden Claw.

After dealing with Arvel, enter the following rooms which contain several Draugr that need to be defeated. While proceeding through this section of the tomb, be wary of any pressure plates. Stepping on these plates will trigger spiked walls to swing.

If you did not kill Arvel, he can be found either near a Draugr or at a spiked wall. Loot his body and retrieve the Golden Claw.

With the claw in hand, continue through the tomb while avoiding any obstacle in your way. The traps within this area will not be too difficult, since this is the game's first dungeon.

Using The Golden Claw

Eventually, you will come to a room with a small water stream running through it. Pull the chain to open the gate at the end of this room, until you are met with a Nordic puzzle door.

At this door, you will need to set the rings to the correct symbol, and then use the Golden Claw as a key. The symbols can be seen on the claw from your inventory. Below, you can check out the correct order of symbols from top to bottom.

- Bear
- Moth

- Owl

Once the symbols are correct, place the claw and open the door.

Obtaining The Dragonstone

On the other side of the door, you will find the first Dragon Shout; Unrelenting Force. This shout will push away those that stand in your way. After learning the shout, a Draugr will rise from a sarcophagus. Kill them and then loot the Dragonstone off his body.

Returning To Farengar

With the Dragonstone in hand, it's time to return to Whiterun and speak with Farengar. This will end the quest, and immediately begin the following main story quest, titled Dragon Rising.

Prior to returning to Farengar, you can stop by Riverwood and complete the Golden Claw quest.

Rewards

As a reward for finishing Bleak Falls Barrow, you will be rewarded with enchanted armor depending on your current level, as well as permission to buy Breezehome; a house available in Whiterun.

DRAGON RISING

How To Start Dragon Rising

The quest begins shortly after bringing the Dragonstone to Farengar Secret-Fire in Dragonsreach at the conclusion of Bleak Falls Barrow. Regardless of what sequence of events led you to collecting the Dragonstone, Jarl Balgruuf's housecarl, Irileth will arrive shortly to deliver the news.

Follow her up the stairs and listen to the conversation. Once the situation is clear, Jarl Balgruuf asks you to join Irileth in stopping the dragon because running in fear from one makes you the most experienced dragon-slayer in Skyrim.

From here, you need to go to the Western Watchtower outside the city. You can follow the quest marker there or join Irileth. It's completely up to you. The Western Watchtower is, obviously, west of Whiterun and easily accessible if you follow the road.

Following Irileth lets you watch her rally a team of Whiterun Guards to take on the dragon. There's no gameplay benefit, but it may be interesting to watch.

How To Fight Mirmulnir

The action starts quickly after you reach the burning watchtower. A panicked guard shouts a warning right as the dragon responsible for the destroyed tower, Mirmulnir, flies in from the mountains. It may be a massive fire-breathing lizard, but it's nothing that you can't handle.

For now, focus on staying close to the watchtower to protect yourself from the Mirmulnir's attacks. His breath attacks will kill you in moments and keeping a stone wall between yourself, and a fiery demise might be the safest tactic.

Mirmulnir breathes either fire or frost in a cone in front of him, dealing significant d

---

Mirmulnir makes a ground-shaking landing that staggers nearby people

---

When grounded, Mirmulnir can attack nearby enemies by biting them or swingin dealing heavy damage.

The most obvious challenge for this fight is that, like the vast majority of dragons, Mirmulnir flies. If you're an archer, hitting him will be challenging but ultimately doable. If you're anything else, it will be a waiting game. Low-level destruction spells struggle to reach flying dragons while melee fighters are just out of luck. That said, Mirmulnir will occasionally land, giving you a chance to fight him up close. Be cautious when attempting this; his melee attacks hit hard, especially at low levels.

Mirmulnir can perform a Finisher on you, killing you outright, if you are below half your health and within biting range. Try to pull back and heal up if things get dicey.

Throughout the fight, remember that you are not alone: Irileth and the guards will be fighting alongside you. Their bows will be able to constantly damage the flying lizard, and their very presence can serve as a distraction. After all, you're the main character of this story, and they don't even have names. It's an easy choice.

Dragons cannot fly after losing half their health. Given enough time, Mirmulnir won't be able to fly away from you.

Once you've slain the beast, something interesting starts to happen.

Finishing Dragon Rising

To the surprise of the many onlookers, you start to absorb the dragon's soul and his knowledge with it. It automatically helps you understand the words you saw at Bleak Falls Barrow, the first word of a Shout. With this, it's official. You're the Dragonborn. One of the guards proposes an optional objective, asking you to demonstrate your power to them. It's a great introduction to your new power and will add some context to who you are.

It may be morbid, but make sure to loot Mirmulnir's bones and scales. They're heavy, but they are worth a lot of money. They're also useful for high-level smithing if you're planning that far ahead.

You may find that you're temporarily unable to fast travel directly to Dragonsreach. Don't worry too much, just walk through Whiterun. You'll soon understand why. Making your way back to the Jarl will start the next quest, The Way Of The Voice.

You will have to speak with him a second time to end Dragon Rising and receive your reward for saving the city, the title of Thane, an enchanted axe, and your personal Housecarl, Lydia. She's a follower that is sworn to help you in your endeavors and to carry your burdens.

You'll see a pair of Alik'r warriors at Whiterun's gate when returning. Talking to them will start the quest In My Time Of Need but won't change much else.

THE WAY OF THE VOICE

How To Go To Ivarstead

The Greybeards live in a particularly inaccessible location. Mountains are hard to climb at the best of times and with their monastery, High Hrothgar, being as high up as it is, it's even harder for you. To reach them, go to Ivarstead, a town at the foot of the mountain.

Broadly speaking, there are three routes to get to Ivarstead.

1.   Head east of Whiterun, travelling past the north face of the Throat of the World until you cross a bridge southeast of Fort Amol. Take a winding path up the mountain into Ivarstead.

2.   Follow the west road from Riften, traveling through the aspen forest to reach Ivarstead.

3.   Take the south path from Helgen, traveling east through a narrow, frozen ravine until you reach the Rift's aspen forests. Turn north to reach Ivarstead.

At the town, look out for a pair of locals talking about climbing the 7,000 steps to High Hrothgar. They're standing on the bridge to the mountain itself, where you can begin the ascent to meet the Greybeards.

One of the locals, Klimmek, is too old to carry supplies up to the monastery. You can offer to do it for him to earn a monetary reward.

How To Reach High Hrothgar

The climb up the steps to High Hrothgar is more challenging than it may appear. While it doesn't require any physical exertion on your part, you'll still be occasionally attacked by wild animals and monsters, depending on your level. Of note, you'll encounter Ice Wolves, which are deadlier wolves, Snow Bears, and Ice

Wraiths. You are guaranteed to encounter a Frost Troll.

The Frost Troll is incredibly deadly to lower-leveled players. The monster has frost resistance, regenerating health, and a mean swing.

If you're struggling to kill it, exploit its weakness to fire and learn to read its attacks to back away and out of range. If all else fails, you can try running past it or finding an alternative route.

Speaking To The Greybeards

The troll is the last hurdle to cross before you make it to High Hrothgar to finally meet the Greybeards. In true mysterious sage fashion, they emerge from the shadows to speak with you. Follow the quest marker to speak to Arngeir, the only Greybeard that will speak without using The Voice.

At his request, use a Dragon Shout on him to prove that you are Dragonborn. Don't worry, he can take it.

This will start a long conversation that gives you the chance to learn more about the history of the Dragonborn. If you wish to progress, select the "I am ready to learn." dialogue option when it comes up.

You only need to explore one dialogue option before you can end the conversation and begin training.

The Greybeards will test you by making you learn a Word of Power, the second word of Unrelenting Force. They carve it into the ground itself. Stand over the carving and look at it to unlock the word. After that, the Greybeards provide you with the understanding needed to use the word without making you use a Dragon Soul.

The trigger for learning the word may not activate when you expect it to. Try to look away and back if you are having trouble.

To test your understanding, they'll summon ghostly versions of themselves for you to test your new shout at. Hold down the shout key to use both words and make sure to hit the phantoms with your shouts to dispel them. Do it three times to end this stage of the quest.

With your skill demonstrated, they'll bring you to the courtyard behind the building for the final test. Here, you'll learn the first word of a new shout, Whirlwind Sprint. Once again, you'll need to learn from carvings on the ground before absorbing their knowledge. Unlike Unrelenting Force, this shout will propel you forward instead of unfortunate victims. Use the shout to pass through the marked gate to pass the trial.

Finally, speak to Arngeir to conclude The Way of the Voice. He will tell you that the next step is to retrieve the horn of Jurgen Windcaller from his tomb. With that, the quest has ended.

THE HORN OF JURGEN WINDCALLER

Getting To Ustengrav

The ancient tomb of Ustengrav is northeast of Morthal, deep in the foggy marshes. If you've never visited, the quickest way is to take a carriage to Morthal and travel directly northeast through the marsh until you find the entrance and a small bandit camp. There will be a necromancer capable of raising fallen bandits with them, so be on your guard for undead. With the path clear, descend into the tomb to begin the trial.

How To Clear Ustengrav

Your first steps into the dungeon will show you a truly unusual sight. Necromancers, bandits, and draugr are engaged in a three-way battle throughout the entrance. The mixture of melee attackers and magic slingers can make the battle ahead challenging.

You can hang back and let the fight play out before mopping up weakened targets.

After the initial set of encounters, you'll be back to fighting draugr through the tomb. The usual tactics will apply here as you fight your way to the Horn.

❖ The Oil Room Puzzle

Ustengrav's first puzzle is quite simple and only requires some time to explore the surrounding area. You'll encounter it in a room with two draugr-filled coffins next to a large puddle of oil. Naturally, make sure to dispatch the enemies before starting.

You're looking for an enchanting table locked behind a pair of metal grates on the left side of the room. A switch by the door will open the inner grate. Look for another one by the coffin on the left side of the room to raise the other grate and access the loot.

As always, this loot is leveled, and your mileage may vary, though it is almost certainly going to be beneficial in some way.

❖ Become Ethereal Word Wall

The dungeon's next area is a large, open cavern filled with skeletons. Be ready for a number of archers among them, making them hard to walk around. Though there's a lot of loot scattered around the room, though the real prize is the Word Wall at the bottom of the dungeon. You'll have to walk down a path at the back of the cavern to reach the Word Wall and unlock a word of the Become Ethereal shout.

Skeletons are fragile relative to the draugr. It won't take much effort to kill them.

How To Solve The Ustengrav Three-Stone Puzzle

Crossing the bridge in the middle of the cavern will bring you to a trio of stones and a doorway blocked by metal grates. This puzzle requires you to step on all three stones and quickly pass through the doorway before the grates crash back down.

There will be a skeleton archer in the puzzle area. You won't need to fight it if you're quick but be aware that there is one.

To succeed, you'll need to make full use of your new Whirlwind Sprint shout. Sprint across all three rocks, shouting right as you activate the last rock. The timing can be tough to get exactly right, but that's all that you'll need to master to complete it.

❖ Getting Through The Fire Traps

You'll be confronted with the next challenge immediately after you dash through the door. Here, you'll need to run across a floor made of flamethrowers activated by pressure plates. Your sprint should be enough to outrun the fire though Whirlwind Sprint will make it much easier to avoid.

If you were burned by the fire traps, the spiders may be especially deadly. You can try breaking through the webbed-up passage at the back of the room to outrun them if you're desperate.

Retrieving The Horn Of Jurgen Windcaller

Finally, with the challenges overcome, it's time for you to retrieve the Horn of Jurgen Windcaller. Walk to the quest marker to discover that someone has stolen the Horn and that they want to meet you in Riverwood to talk.

There are many treasure chests on the way out of Ustengrav. Make sure to loot them.

Unfortunately, you don't have much of a choice. Return to that town and look for Delphine to ask for the attic room. This will start the quest "A Blade In The Dark" and will see the Horn returned to you in short order.

With the Horn in tow, return to the Greybeards to complete the quest and receive the final word for Unrelenting Force.

A BLADE IN THE DARK

Starting A Blade In The Dark

It starts at the tail-end of The Horn of Jurgen Windcaller, the previous main story quest. Delphine will force you to go through some song-and-dance to appease her paranoia, culminating in her bringing you through a staircase hidden in her closet. There, she will tell you that dragons are being resurrected in Skyrim. The rest of the conversation is more exposition.

Delphine's identity may not be such a big revelation if you've been paying attention She was the woman talking to Farengar at the end of Bleak Falls Barrow.

Delphine asks you to join her in killing a dragon to prove that you're the Dragonborn, having identified Kynesgrove as a likely location for the next resurrection. You can travel with her or simply meet her there. Kynesgrove itself is closest to the city of Windhelm, allowing you to quickly reach it after a carriage ride to the city. Simply follow the road south. A frightened resident will tell you about the dragon when you arrive.

Following Delphine will let you learn more about her as she speaks occasionally while you journey, though it is a lengthy trip. You should find your own way there if you aren't interested.

Fighting Sahloknir

When you reach Kynesgrove, you'll find the villagers fleeing the town in panic. Run up the hill, following the road to reach the dragon mound where you will find the black dragon from the beginning of the game speaking. You won't be able to do anything save for listening to the two dragons. After a lengthy exchange in their own language, you'll have to fight the freshly resurrected dragon, Sahloknir. You will have to defeat him to proceed.

It may be the Skyrim equivalent of a sucker punch, but you can attack him before he fully resurrects, letting you get in some damage before he can retaliate.

Like most dragons, Sahloknir will take to the skies and breath either frost or fire at you. Unfortunately, the dragon mound itself offers scant cover. The best shelter from his breath will be a handful of rocks by the road.

Sahloknir isn't a particularly unique dragon and will be defeated the same way you've defeated other dragons. Fire arrows at him and attack him with your other options as soon as he gets into range. Pummel him into submission and consume his soul at the end of the fight. With that out of the way, Delphine will be forced to recognize who you are. She tells you to meet her in Riverwood, starting the next quest.

DIPLOMATIC IMMUNITY

Unlocking Diplomatic Immunity

This quest follows after A Blade In The Dark, a quest that had Delphine demand you kill a dragon to prove to her that you are Dragonborn. At its conclusion, Delphine will trust you enough to reveal her Blade identity, as well as give you the key to her secret room underneath Riverwood's Sleeping Giant Inn.

Delphine's Plan

Safely secured in Delphine's secret office, she'll bring you in on her plan for finding out what the Thalmor knows about the return of the dragons. Thalmor Ambassador Elenwen loves throwing lavish parties at the Thalmor Embassy, so Delphine plans to send you to a party, disguised as an important guest.

Once you're inside, you'll search for any and all information the Thalmor has hidden away. It's a classic spy mission, and Delphine has someone on the inside: a Bosmer Elf named Malborn.

Meeting Malborn

Delphine will send you to the Winking Skeever tavern in Solitude in order to meet Malborn, who can be found relaxing inside. He works at the Thalmor Embassy and offers you a chance to sneak in items that wouldn't be allowed past security. You'll only get one chance to give Malborn items, so make sure you have everything you want to give him when you talk to him.

Weapons	Naturally, you will want to give Malborn your weapons of choice for this infiltration. While bringing your favorite stabbing implement would be obvious, don't forget to bring weapons that you don't normally use but find useful. For example, if you find yourself falling back to your bow when enemies get out of reach, don't forget the bow and sufficient arrows.
Armor	Delphine will make you don a disguise before you enter the Embassy, preventing you from rocking up in full plate armor. Make sure to pass Malborn your protective equipment like your armor and shields
Potions	Your stockpile of Potions of Healing, Magicka and Stamina won't be added automatically to the hoard. You will need to make sure that you'll be fully equipped for the task at hand.
Miscellaneous Items	Other items such as lockpicks may come in handy for infiltrating the embassy. Make sure you bring along any other aids you may use such as food, if necessary.

Malborn has no limit to what he can smuggle in, but make sure it doesn't over-encumber your character later.

If you've finished The Companions quests, you may wish to bring Wuuthrad with you. There are few instances where you are guaranteed to be able to take full advantage of its bonus damage to elves.

Entering The Party

After you give your desired equipment to Malborn, you'll head to Katla's Farm in Solitude where Delphine will give you both your official party invitation and proper Party Clothes so the Dragonborn will blend in. Delphine also will hold all the gear and items you didn't give to Malborn, which you can retrieve in her office in the Sleeping Giant Inn afterward.

You'll automatically board the carriage heading to the Embassy after talking to Delphine, so make sure you're ready when you speak to her.

Arriving at the Embassy, you'll meet the drunk Razelan of the East Empire Company (who will be very useful later on), as well as a Thalmor Guard checking your invitation. Elenwen will ambush you immediately after entering, wanting to know everything about the mysterious guest at her party. Luckily, your friend on the inside, Malborn, will distract her for us before she sees through the disguise. Grab a drink from Malborn afterward and dive into the party and speak to the guests.

Creating A Distraction

Players who have completed many quests will find themselves at an advantage when it comes to creating a distraction. Certain characters may appear depending on which quests you've fulfilled. Almost all of them can be used to create a distraction and allow the Dragonborn to slip away unnoticed. Here's a full list of who you can encounter:

Balgruuf the Greater	Players who have finished the Battle for Whiterun will run into Balgruuf at the party, who is distinctly not having a good time in the presence of Thalmor. The Jarl will be glad to help you out if you ask, promptly starting a fake argument with the inebriated Razelan.
Elisif the Fair	Elisif's appearance is randomized, but whether or not she'll help you out depends on if you've completed Elisif's Tribute, a quest that involves honoring her deceased husband, former High King Torygg. If you have, you'll need a Speech Skill of at least 50 to get her help, which comes in the form of bullying Razelan.
Erikur	This thane is another character with a randomized appearance, but getting his help only requires a little manipulation... and

	messing with a woman's career. Erikur spends the party being sleazy towards one of the servers, a Bosmer Elf named Brelas. The Dragonborn can speak to Brelas about this unwanted attention and use it to create a distraction by talking to Erikur again. No matter if you lie or tell the truth about Brelas, she will be thrown in the dungeon, as it's Erikur's word against hers. Thankfully, you can free her later in the quest, the least you can do after putting her through such trauma. After being freed, Brelas moves to Windhelm, living in the Candlehearth Hall.
General Tullius	Highly regarded Imperial General for the Legion, Tullius appears if the player has finished Message to Whiterun and there is no civil war attack ongoing. After all, a war General would have more important things to do than attend a Thalmor party. You won't get any help with distractions from him, however. He's just there to vibe.
Idgrod Ravencrone	The Jarl of Hjaalmarch is a fascinating woman who uses "visions" from the Eight Divines to guide her decision-making. If you've completed the quest Laid to Rest and are not currently on the quest Thane of Hjaalmarch, Idgrod will appreciate your previous aid and will also cause a ruckus for you. She envisions terrible things about Razelan, the poor man, though she means him no harm.
Igmund	The Jarl of Markarth has a random chance of appearing, and all you need to enlist his help is a relationship above zero (meaning you've assisted him before) and a Speech Skill of 50 or above.
Maven Black-Briar	You'll most likely see Maven Black-Briar at the party unless she's the Jarl of Riften or you're on the quest Dampened Spirits. Having a positive relationship with her will convince her to help you out, but you'll have to be the Master of the Thieves Guild to do it.
Ondolemar	One of the most interesting interactions in the entire quest is getting an Altmer Elf to help you steal Thalmor information, even if it is unintentional. Ondolemar is a Thalmor Justiciar in the Understone Keep of Markarth, and players who have helped him expose Bard Ogmund's Talos Worship (in the quest Search

	and Seizure) can request a favor from him. If Markarth is still under Imperial Rule and you have a Speech Skill of 50 or above, Ondolemar will agree and start harassing Razelan.
Orthus Endario	Orthus is a pretty average guy compared to the rest of the guests. In order for Orthus to appear and help you out, you'll need to have completed the Rise in the East quest, which involves slaying some pirates for the East Empire Company.
Proventus Avenicci	Though Proventus can't help with a distraction, he can appear as a guest if Message to Whiterun is not currently active and The Whispering Door is either complete or below stage 80.
Razelan	The reliable drunk, Razelan will always be at the party and is used in almost every character's distraction method, but he's also the easiest distraction option himself, especially for players who can't convince others to help them. Simply give Razelan a drink and he'll go off on an inappropriate speech about Elenwen, giving you time to sneak off.
Siddgeir	The Jarl of Falkreath may grace the party, but only a Thane of Falkreath with a high Speech Skill could possibly convince him to assist.
Vittoria Vici	Famous Solitude bride-to-be has a small chance of appearing at the party unless the Dragonborn has started/completed the Dark Brotherhood quest Bound Until Death. If players have completed The Spiced Wine quest and gotten Vittoria to relinquish Evette San's Spiced Wine from costly tariffs, they'll have the option to persuade her to create a distraction (solely as a prank, of course). Like other guests, the Dragonborn will need a Speech Skill of at least 50 to convince her.

While the incident is occurring, simply sneak out of the main room and towards the kitchen.

The Sneaking Begins

Once in the kitchen, you'll notice Malborn has blackmailed the Khajhit cook not to say anything about your presence, so you can safely grab your stashed items and proceed to the next stage of the quest, as Malborn locks the door behind you.

You'll then need to formulate a strategy for getting past all the Thalmor guards. Wearing Hooded Thalmor Robes can help make it harder to detect you, but you'll have to encounter the Thalmor Wizard upstairs. Most players opt to practice their stealthy sneaking around the guards' patrols. If you keep getting caught, murder is also an option.

If you are a High Elf, wearing Hooded Thalmor Robes will allow you to pass the guards so long as you keep your distance.

There are guards having a conversation in the first hallway you'll go through, which you can listen in on. Finishing their conversation allows you time to sneak past them upstairs and kill the Thalmor Wizard quietly and get his robes.

The outside courtyard is tricky, considering there are three different Thalmor guards and a Wizard who won't buy your disguise. You'll need to enter the Solar of Elenwen, though, so attacking them may be necessary.

Stealth archery can help you gain an advantage and not be immediately swarmed by guards. Spells like Calm and Fury will also work in your favor, as well as invisibility and high Stealth.

Extracting The Info

Once in the Solar, find the plant to hide behind and listen to the talk between Thalmor Rulindil and informant Gissur. They discuss their interrogation victim, Etienne, along with an unnamed old man. Gissur will leave the building after and Rulindil goes to the dungeon, leaving you to sneak past (or kill) a guard to get the interrogation chamber key from a chest in the northwestern corner.

If you're not a milk-drinker, you can save some time by killing Rulindil before he goes to the dungeon, taking the key from his corpse instead. Doing so will likely involve abandoning any pretense of stealth.

Also in the chest are files by the Thalmor on Delphine, Ulfric Stormcloak, and their investigations on the Dragons, exactly the info Delphine wanted us to find.

In the dungeon, Rulindil will be making a guard torture Etienne for info until he passes out. You can then proceed to kill both of them to get what you need, or you can use stealth to sneak past them both. Fighting Rulindil is not recommended, due to his powerful wizardry skills, though he does have considerable loot to grab. A stealthy backstab would work nicely.

Making Your Escape

You can free Etienne and ask him for information now, or you can get the same info from the Thalmor Dossier: Esbern in the chest near his cell. After this, go up the east staircase and you'll find Malborn, who has been revealed as a traitor. The two Thalmor soldiers will kill him on the spot if you don't intervene, so make sure to act fast or save beforehand. The soldiers will have a trap door key on them, making it that more reasonable to kill them.

If you used Erikur to provide a distraction earlier, this will be the only moment you can free Brelas, who was thrown in the dungeon for the incident.

You now have NPCs following you, none of which can fight or defend themselves. Swell! Use the trap door key to enter Reeking Cave and leave the Thalmor Embassy behind for good.

Your inept, temporary followers will try to kill the troll waiting in the cave. This will most likely lead to their death, so try to run ahead and take care of it if you want them to survive.

Notable items in Reeking Cave include the Illusion Skill book Before the Ages of Man and an Unusual Gem, which is one of twenty collectibles needed for No Stone Unturned.

Once you've exited the cave, your followers will all disperse and go their separate ways, so you can head back to Riverwood and inform Delphine of what you learned as well as grab the rest of your equipment. Telling her the Thalmor want Esbern shocks her, as Esbern is a former Blade living in hiding in Riften. This will be your next destination in the journey, but this marks the completion of Diplomatic Immunity. A Cornered Rat will start right after. A Dragonborn's work never ends!

A CORNERED RAT

Starting A Cornered Rat

The quest begins at the conclusion of Diplomatic Immunity after Delphine learn that the Thalmor are looking for Esbern. Figuring out that the most likely place for a man to go into hiding is the Ratway in Riften, she asks you to look for Brynjolf.

If you are already a member of the Thieves Guild, this section becomes much easier

Finding Where Esbern Is Hiding

Being a Hold Capital, you can easily reach Riften by taking a carriage directly to it. There, look for the city's inn, The Bee And The Barb to find Brynjolf. When you arrive, there are a few ways to find out where Esbern is.

Talk to Brynjolf	Assist him: He will demand that you help him with a scheme before he gives you any information. This will start the quest A Chance Arrangement, the first quest in the Thieves Guild questline.

	Persuade him: You can tell him that you are looking for Esbern for the good of Skyrim. If you have a sufficiently silver tongue, he will accept that it makes sense and tell you where Esbern is.
Talk to Keerava	Asking the Argonian barmaid Keerava for suggestions will have her point the way to the Ratway, though she can't offer any further suggestions. When you arrive, you will need to learn where Esbern is from Vekel the Man.
Ask as a member of the Theives Guild	If you've already become a member of the Ratway, you can simply ask Brynjolf for Esbern's location.

Reaching Esbern

Getting to the old hermit will need you to get through the Ratway, a collection of tunnels underneath Riften where many of its worst denizens live. On your first visit you will need to clear it of its violent inhabitants.

We covered how to get through the Ratway in our guide for the quest Taking Care Of Business.

When you reach The Ragged Flagon, take the doorway to Ratway Tunnels, located by Vekel's bar. From here, you will be fighting Thalmor agents mixed with more Ratway residents. You will need to descend through the tunnels, taking a right on the middle floor to find the way to the Ratway Warrens where you will find Esbern.

Esbern will be easy to find at the entrance. Simply reach the open area in the center and take the stairs up to the bolted-up door. Interact with it to speak to Esbern, though how it proceeds from here depends on you.

- Persuade: Dragonborn with a sufficiently high speech skill will be able to convince him to let out in
- Use the passphrase: Asking him what happened on the 30th of Frostfall will automatically convince him.
- Tell him you are the Dragonborn: He will be convinced enough to let you in

On entering and after a brief conversation, the quest ends and the next one, Alduin's Wall begins.

Alduin's Wall will start immediately. There won't be any downtime for a while.

ALDUIN'S WALL

Starting Alduin's Wall

This quest automatically starts at the conclusion of A Cornered Rat. You won't have the opportunity to head off and do a different quest before this one begins. Esbern needs to be escorted to Riverwood immediately.

Fortunately, Esbern is marked as essential and cannot permanently die.

You will be attacked by a powerful Thalmor Wizard immediately after leaving his hovel. The wizard will attack you with lightning spells and is able to summon a Frost Atronach, an enemy that excels at fighting in melee. It will be a difficult fight, though Esbern will offer some assistance.

Many of Esbern's violent neighbours may join in the fight, disrupting it and making it more dangerous.

From there, retrace your steps back to the Ratway Tunnels, crossing the open area to find a staircase back up to the Ragged Flagon and eventually back into Riften. You can fast travel directly to Riverwood from here.

At the town, bring Esbern into the Sleeping Giant Inn to reunite him with his old colleague, Delphine. In a meeting at Delphine's secret basement, he shares what he knows. The ancient Blades have recorded the secret to defeating Alduin in Sky Haven Temple, an old base of theirs. You need to find the ancient ruin to discover that secret.

Reaching Karthspire And Sky Haven Temple

The two Blades will start following you to Karthspire, a Forsworn camp built into Sky Haven Temple. The closest city is Markarth, letting you take a carriage ride there if there are no nearby fast travel options.

If you wish to complete other quests before continuing, you can tell the Blades to meet you there. You will find them near Karthspire when you are ready to continue.

Karthspire is home to a large group of Forsworn, a group of rebels that lament the loss of their native land. You won't need to kill all of them, though they will be immediately hostile. They are far more formidable than ordinary bandits, using unique and hard-hitting weapons alongside magic from their own shamans and hagravens.

There is a very high likelihood of a dragon attack. Its presence will make the fight more chaotic, though it could be a useful distraction,

You will need to enter Karthspire itself through a cave on the west side to reach Sky Haven Temple. Inside, you will need to clear the Forsworn blocking you from reaching Sky Haven Temple itself. There will be a powerful Forsworn Briarheart among them who will devastate unprepared players.

Immediately after, you will need to complete a series of puzzles to continue.

How To Solve The Sky Haven Temple Rotating Pillars Puzzle

The path will be blocked by a raised stone bridge. To lower it, complete the first puzzle. It consists of a trio of rotating pillars with a unique set of symbols on them. Esbern will offer hints while commenting on the history of the location. To solve it, rotate all the pillars to the circular symbol marked with a downwards arrow. The bridge will fall and you will be able to proceed.

How To Solve The Sky Haven Temple Fire Floor Puzzle

In the next room, you will need to cross a room that has a floor made of pressure plate-activated flamethrowers. This puzzle is similar to the one in Ustengrav encountered in The Horn Of Jurgen Windcaller.

Your companions ask that you deactivate the trap for them to pass safely. To do so, pull the chain mounted on the obelisk on the other side of the room. Using Whirlwind Sprint will let you reach it without being caught in the trap.

You can also sprint across without shouting if you are feeling impatient.

Pulling the chain will also drop the stone bridge, letting you go to the final puzzle.

Entering Sky Haven Temple

At the very end, you find yourself in a chamber sporting a large, bald head with no clear way forward. Wait for Esbern to guess how to open the way forward before following his advice, interacting with the stone circle in the center of the chamber, and spilling your blood onto it.

With the way open, enter Sky Haven Temple and explore. You will need to wait for the Blades to discuss Alduin's Wall for the quest to progress. They'll realize that they need the Graybeards to discover what the technique is, starting the next quest The Throat Of The World

THE THROAT OF THE WORLD

Starting The Throat Of The World

This quest starts right at the end of Alduin's Wall, as Delphine struggles to guess who might know a shout powerful enough to defeat the first dragon. You suggest the Graybeards, the hermits that taught you the words for Unrelenting Force and set you on the path to retrieve the Horn of Jurgen Windcaller.

Returning to High Hrothgar, you look for Arngeir in a bid to discover this legendary Shout.

Ascending The Throat Of The World

Though the old sage has his reservations, his fellow Graybeards remind him that denying you the knowledge is not his to make. That decision lies with Paarthurnax, a wise being living at the very peak of the mountain.

Gathering in the courtyard behind the monastery, the Graybeards teach you three words of a wholly new shout, Clear Skies. As they explain, the path to the summit has been blocked by an eternal blizzard. Attempting to walk through will hurt and quickly kill you.

Clear Skies can be used outside of the quest, letting you manipulate the weather into a sunny day.

While it has no direct gameplay benefits, bending the weather to your will is always a trip.

Clear Skies will be able to clear the blizzard for a short time. You will have to use it to open a path as you climb up the mountain by removing sections of blizzard as you advance.

Along the way, you will encounter ice wraiths hidden in the sleet. These enemies can be hard to spot, especially if you are making the ascent during the day. Their damage, however, is not hard to miss.

If you are struggling to defeat the ice wraiths, use fire against them, even the Flames spell works wonders.

The way up the mountain isn't too long. With sufficient uses of Clear Skies, you'll eventually make it to the summit and the mysterious hermit at its peak.

Meeting Paarthurnax

The hermit that the Greybeards look up to is none other than an ancient dragon. As he explains, he took up residence high above Skyrim to watch for the return of his brother, Alduin.

He will demonstrate the shout Fire Breath, allowing you to learn a word and requires you to demonstrate your prowess back to him by shouting back at him, dispelling any doubt as to your identity.

A unique weapon, the Notched Pickaxe, is at the very peak of the mountain.

With formalities settled, Paarthurnax will explain how you can learn the shout that defeated Alduin; by using an Elder Scroll. This starts the next quest, Elder Knowledge.

ELDER KNOWLEDGE

Starting Elder Knowldege

This quest begins after your introduction with Paarthurnax. He tells you that the only way to see and learn the shout used to defeat Alduin is to read an Elder Scroll at the Throat Of The World. Unforuntately, being a hermit, he does not know where to find one and advises you to speak to your allies to find out.

You may speak to Esbern or Arngeir to find out, though they will give you the same answer. The College of Winterhold would be the most likely people to know.

Searching The College Of Winterhold

Entering the College isn't as straightforward as walking in. A High Elf, Faralda, will demand that you prove your worth to the College before she lets you in. Here, you have three options.

Tell her you are Dragonborn	Demonstrating the Thu'um to her is impressive enough for her to let you in.
Take a test	Proving that you have a degree of magical proficiency will demonstrate that you are skilled enough to enter.
Pursuade	You need to have Speech 100 for this to succeed, though you can convince her to step aside.

Entering the College will also start the quest First Lessons, the first part of The College of Winterhold questline. You do not have to play the rest of the quest to progress, however.

You want to visit the Arcanaeum, the College's library, to look for clues. Asking the librarian, Urag gro-Shub, for help will get you a pair of books on the topic. One of them, Ruminations on the Elder Scrolls is genuinely impossible to understand. Bringing it up with Urag will tell you about Septimus Signus, an expert on Elder Scrolls that happens to be alive.

Reading Ruminations on the Elder Scrolls will also begin the quest Discerning The Transmundane, large parts of which runs in parallel with Elder Knowledge.

Finding Septimus Signus

With your only lead identified, you need to head into the Sea of Ghosts in the north of Skyrim to find Septimus. He will be found far to the north in the middle of ice floes at Septimus Signus's Outpost.

What he says is incredibly hard to follow but you decipher his ramblings regardless You can find an Elder Scroll in old Dwarven ruins buried beneath Skyrim. Giving you a Lexicon and an attunement sphere, he points you at Alftand, an exposed ruin that you can use to enter the underground Dwarven city.

The next part of the quest involves and extended dungeon delve. Make sure that you are fully stocked on potions and other supplies before you begin.

Getting Through Alftand

Alftand itself is in the middle of Winderhold. Though getting there is hard, the ruins themselves will be hard to miss. Climb on to the exposed bridges and find your way to the marked entrance. You will be able to enter the Alftand Glacial Ruins from there. Inside, you will find the remnants of a recent expedition.

One member, J'darr is alive and has gone insane from Skooma withdrawals. He will attack you immediately though he won't be hard to kill.

Follow the trail of dropped torches and lanterns to navigate through Alftand. Dwarven Spiders will attack you as you explore, recognizing you as a trespasser. If this is your first time fighting them, their small stature and explosive death may catch you by surprise.

That said, they are far less dangerous than Dwarven Spheres. Their presence marks a sharp increase in difficulty as the old constructs can easily defeat low-leveled players.

Dwarven Spheres are resistant to magic and immune to frost and poison damage. They are also able to fight you up close and from afar.

You will have to fight through several rooms filled with Dwarven constructs until you reach Alftand Animonculory.

Alftand Animonculory

The next leg of the journey gets more complicated. One of the first challenges you encounter is a Dwarven Sphere in an oil-filled room. Setting the oil on fire will damage it and give you an edge as you go deeper.

A ramp covered in pressure plates will block your path through the ruins. Triggering it will send a pair of spinning blades down there ramp, dealing massive damage. Use Whirlwind Sprint or try and dash up in a zig-zag pattern to avoid the pressure plates.

You can find clues about the fate of the expedition if you pay attention to the corpses you find in this area.

With that trap out of the way, you will need to descend deeper into Alftand. Reaching a large, deep shaft, you need to navigate fallen masonry to reach the bottom and exit. You will encounter Falmer as you descend further down the mine shaft. These enemies are more deadly than your average bandit and will put up a tough fight.

The Falmer make frequent use of poisons. Having a countermeasure for lingering damage over time will help you survive the encounters.

You will reach a Falmer camp at the bottom. Be prepared for more of the blind enemies waiting for you. It doesn't get better in the next room, where Falmer archers use flaming machinery to keep you at bay while shooting with impunity.

The Falmer are blind, making them easier to sneak around or shoot from beyond their hearing range. They are capable at detecting you from a distance, just that they use their hearing instead.

Muffling your sound is a better strategy against them than going invisible as a result.

The two Falmer-filled rooms mark the end of the area and bring you to Alftand Catherdral

Alftand Catherdral

The final part of Alftand, though far from the last part of your journey, is the Cathedral. This area will have you fighting more Falmer and even a Dwarven Centurion.

It will be a direct path through this area until you reach a gate. There, take the stairs by the entrance to the room and pull a lever. This will open the gate and awaken a Centurion.

Dwarven Centurions are slow, lumbering enemies that excel at fighting in close quarters. Their sole ranged attack is spewing steam breath at enemies though their resistance to ranged attacks limits the effectiveness of staying beyond its reach.

After dispatching the Centurion, you will come across a pair of living humans by a pedestal. As the remaining survivors of the expedition, they start bickering before turning hostile. You can leave them to kill each other or join in.

Umana, one of the humans, has the unique shield The Targe Of The Blooded, a spiked shield that makes enemies bleed when bashed.

Interact with the pedestal to insert the attunement sphere to progress further into Blackreach.

Navigating through Blackreach

Blackreach is a massive, underground area that forms a whole new part of Skyrim. There is a lot to this place but to complete Elder Knowledge, you need to focus on reaching the Tower of Mzark.

Take a left immediately after leaving Alftand to reach a Dwarven Centurion head with a button. This will activate a lift that will make it easier for you to get to Blackreach in the future.

To do that, head west toward the quest objective, fighting the many enemies that will get in your way. Between Falmer, Chaurus and Dwarven constructs, the journey to the Elder Scroll will be difficult.

At the objective, enter The Tower Of Mzark

Solving The Tower Of Mzark Puzzle

At the final step, you find yourself in a massive Dwarven chamber full of refracting lenses. After placing the lexicon in the pedestal, you need to manipulate the mechanism to release the Elder Scroll. To solve it, refract the light into the relevant lenses. If you are struggling, follow the below order. The buttons are numbered from left to right.

This will release the Elder Scroll. Pick it up to end the quest and start the next one, Alduin's Bane.

ALDUIN'S BANE

Starting Alduin's Bane

This quest automatically begins when you retrieve the Elder Scroll from its resting place at the Tower of Mzark at the end of the last quest, Elder Knowledge. Continuing with Paarthurnax's request, you return to The Throat Of The World. After a short exchange with the old dragon, stand on the spot with visible magical distortion.

You can read the Elder Scroll whenever you want. However, doing so will momentarily blind you with no benefit.

Search for the Elder Scroll in the Books section of the inventory and select it to begin reading. This will begin a flashback sequence where you will watch ancient Nord heroes defeat Alduin at the very spot you are standing on.

There is nothing you can do during the flashback except for waiting for the heroes to reveal the secret. As Alduin appears in the flashback, they use the Shout Dragonrend, forcing him to land. Unfortunately, they fail to defeat him and use an Elder Scroll to banish him forward in time.

Fighting Alduin On The Throat Of The World

Not wanting to let you learn the forbidden Shout so easily, Alduin comes to fight you. After some villainous taunting, he will begin attacking.

At Paarthurnax's urging, use Dragonrend on Alduin to force him to land and render him vulnerable to damage. From here, he fights similarly to the dragons you've become familiar with. He uses the following attacks

Bite	Alduin lunges forward to bite at you when you are too close.
Fire Breath	Alduin breathes fire breath, damaging you over time and inflicting lingering damage
Frost Breath	Alduin breathes frost breath, damaging you over time and sapping your stamina
Dragon Storm Call	Alduin calls a meteor storm. Firey rocks will rain from the sky damaging you and inflicting fire damage if you are caught in the impact.

You need to deplete all of Alduin's health to force him to retreat. Your usual dragon-fighting tactics will work fine here, just make sure to keep him under the effect of Dragonrend. He will not take damage when the shout is not affecting him. Paarthurnax will join the fight on your side, attacking Alduin frequently.

Bested, he flees to Sovngarde, far beyond the reach of any living person. Needing to devise a plan, you the next quest, The Fallen, begins.

Using Dragonrend quickly and frequently will stop Alduin from casting Dragon Storm Call, denying him a large source of damage and making the fight easier.

PAARTHURNAX

Starting Paarthurnax

Speaking to either Esbern or Delphine after Alduin's Bane will start the quest. The two Blades, having discovered that Paarthurnax was responsible for atrocities committed in a previous conflict, demand that you kill him.

Considering how the dragon has treated you, killing him may seem distasteful. You can speak to Paarthurnax about the Blades' demand if you are struggling to decide.

This quest will disappear from your quest log if you complete Dragonslayer without killing Paarthurnax.

This will not stop you from being able to kill him. Doing so will allow you to speak to Delphine and make the Blades friendly again.

Choosing Between The Greybeards And The Blades

The hardest part of this quest is deciding which of the two sides to back. While there are plenty of narrative reasons to choose a side, there are plenty of gameplay reasons to choose.

Killing Paarthurnax will align you with the Blades while refusing to do so will align you with the Greybeards.

The side that you do not support will refuse to assist you further, locking you out of their questline and any benefits that they might offer.

❖ *Greybeards*

The Greybeards provide a repeatable quest that brings you to word walls throughout Skyrim, letting you expand your arsenal of shouts with much less guesswork.

Aligning with the Greybeards will keep Paarthurnax alive, letting you choose between the following buffs from him.

Force Without Effort	Increases stagger inflicted to enemies by 25% and resistance to it by 25%
Ethereal Spirit	Increases health regeneration by 25% when using the Become Ethereal shout
The Fire Within	Increases damage done by the Fire Breath shout by 25%

❖ *Blades*

Siding with the Blades lets you continue down the path of being the greatest dragon killer of all. You will be able to send followers to serve as recruits with the Blades.

Once you recruit three Blades, they will assist you in dragon hunts.

Esbern will help you locate dragons to kill, netting you more souls and dragon parts. He will also provide you with more bonuses to fighting dragons by reducing damage taken and increasing your critical chance against them.

Killing Paarthurnax

Doing so is a relatively simple matter. Go to the Throat of the World and attack Paarthurnax. It will take a few hits, but he will eventually turn hostile.

Paarthurnax will take to the skies and begin to fight you like most other dragons. He is no different from any other dragon and will be easily slain with the same tactics.

SEASON UNENDING

Starting Season Unending

This quest starts during The Fallen if the Civil War is still ongoing. The Jarl of Whiterun will not allow you to use Dragonsreach's ancient mechanisms to trap the dragon if there is the risk of the opposing faction attacking while his men are occupied.

He suggests asking the Graybeards for help. Though it takes some persuading, they will allow you to host ceasefire talks at High Hrothgar. After convincing General Tullius and Ulfric Stormcloak to attend, either through charisma or with some arguing, the two sides meet at the monastery.

Negotiating With The Factions

This quest is entirely dialogue-based, with each faction demanding concessions from the other. The decision to settle each one will fall to you over the course of three disputes. While no faction would leave the table, the decisions you make here will affect the Civil War.

The exact contents of the discussion depend on which side you have chosen to fight for, if at all. Neutral players have the same experience as Imperial ones. Where relevant, this guide will be written in generic terms.

❖ Elenwen

The first dispute comes from Ulfric before he even takes a seat. Seeing the Thalmor Embassador, Elenwen, present, he demands that the High Elf leave the conference. Tullius will refuse, pointing out that Ulfric has no right to choose who can attend. Given that the nation she represents forced the treaty that led to the Civil War, Ulfric will not back down on it. You will decide if she stays or goes.

Side With Tullius	Elenwen stays, to Ulfric's dismay
Side With Ulfric	Elenwen leaves, to Tullius' dismay

The talks will continue regardless of your decision.

❖ Trading Cities

With the negotiations underway and representatives posturing, the next stage will have the opposing faction demand a walled City. You will not be able to prevent them from claiming it. However, you can decide how much your side gets.

Trade A Major Hold	You trade a major hold containing a walled city in exchange for the demanded one. This will be seen as a fair trade.
Trade A Minor Hold	You trade a minor hold that has a small city in exchange for the demanded one. This will be seen as an unfair trade favoring your opposing faction.

If any side feels that they have been cheated, they will threaten to walk out. Esbern will make a speech reminding the participants that the fate of the world is on the line, starting the next phase.

❖ Making Consessions

In this final stage, the dissatisfied side will demand compensation for a massacre or demand ownership of a minor hold to be satisfied. You can repeatedly refuse until they give up, or you can appease them. With everything settled, a peace treaty will be signed.

Consequences For Season Unending

Whatever trades were made during the quest will need to be resolved when the Civil War resumes. The conflict will not continue until the main quest is resolved. The Blades will also approach you to start the quest Paarthurnax.

THE FALLEN

Starting The Fallen

A stone castle is perched on top of a mountain, showing a grand hall overlooking surrounding plains

This quest starts immediately after Alduin's Bane ends. You will be left to turn to your three main allies for help:

1. Paarthurnax, who is with you at the Throat of the World

2. Arngeir, with the Graybeards at High Hrothgar

3. Esbern, with the Blades at Sky Haven Temple. Speaking to him will also start the quest Paarthurnax.

All three of them will suggest capturing one of Alduin's allies at Dragonsreach, Whiterun's Keep, and a place that you've spent plenty of time at. You just need to sell the idea to the Jarl.

Convincing The Jarl Of Whiterun

Though initially in disbelief, the Jarl will hear out your request. He will not be happy with the idea and will not agree unless he believes that his Hold will be safe. What happens here will depend on the state of the Skyrim Civil War.

The Civil War has yet to be resolved.	The Jarl will refuse, pointing to the danger posed by his enemies noticing his troops relocating to deal with the dragon. He will need both sides to agree to a truce before he will help, starting Season Unending.
The Civil War has been resolved or is on the verge of resolution.	If your side has won the Civil War or is besieging the enemy Capital, the Jarl will agree to help with your plan, letting the quest continue.

You will not be able to start Season Unending if you have finished Paarthurnax.

The Civil War will have to be brought to a conclusion before you can continue with the quest in this instance.

Regardless of how you get the Jarl to agree, you will need to learn how to call out one of Alduin's allies. You will be sent to Esbern or Paarthurnax depending on how the quest progresses. Doing so will teach you the shout Call Dragon which will be instrumental to the plan.

Capturing Odahviing

A man challenges a dragon inside a stone castle with patches of ground burning

When all the pieces are in place, you can speak to the Jarl to set the plan in motion. Head to the balcony behind the throne room to find the trap. Once you're ready, use Call Dragon to challenge this dragon, Odahviing, to a fight. He will swoop down, killing a Whiterun Guard and starting the battle.

You must use all three words of Call Dragon to challenge Odahviing.

Use Dragonrend on Odahviing to force him to land. Lure him deep into Dragonsreach until he is under the wooden mechanism by the stairs. The trap will fall onto his neck as it automatically triggers.

The objective of the fight is not to kill Odahviing, but to lure him into the trap. There is no need to fight too hard against him.

Odahviing will realize that he has been bested and admits that many dragons have begun to question Alduin's leadership. Your prowess in combat has swayed him to your side, and he agrees to carry you to Skuldafn, an ancient ruin that holds a portal to Sovngarde. This starts the quest The World-Eater's Eyrie.

Starting The World-Eater's Eyrie

The quest follows The Fallen, where you trap Odahviing and convince him to lend you his aid. He tells you that Alduin has retreated to Sovngarde, the Nordic afterlife, to feast on the souls of heroes. As a living person, following him there is normally impossible.

Odahviing explains that Alduin uses a portal at the ancient ruin Skuldafn to reach Sovngarde. As a flightless bipedal Dragonborn, reaching it yourself is beyond you. The only being that can help is Odahviing, who conveniently needs to be released to do so. With no other options, you must agree to his offer to continue.

Talk to the Whiterun Guard up the stairs to release the dragon. Odahviing will stand at the edge of Dragonsreach, waiting for you to get ready. Once you are satisfied, tell him and fly off to the end.

Not only is Skuldafn long and challenging, but you will not be able to return to Skyrim until you have completed the remaining quests. Be aware that your followers will not be able to come with you. Make sure that you're stocked up on potions and other supplies before you get going.

Getting Into Skuldafn

Odahviing will depart after dropping you off, leaving you alone with a whole dungeon ahead of you. As soon as you start making your way forward, you will be attacked by draugr and a dragon.

There are a pair of optional towers containing loot and draugr. You can only reach the southern tower.

After the first bridge, take a right and turn left into a courtyard. You will be attacked by more draugr, likely accompanied by another dragon. Go straight from the courtyard, passing under a pair of bridges, and turn left to go up a staircase to reach Skuldafn Temple.

Skuldafn Temple

A man fights undead warriors in a dusty stone crypt with a puzzle in the background

Inside the ruin, you will fight through a handful of draugr before reaching your first puzzle, a pair of closed-off doorways and a trio of spinning stone pillars.

❖ Solving the first Skuldafn Puzzle

To solve the puzzle, look for four hints spread around the room. These are carvings that match the faces on the pillars. If you face north, they are arrayed in the following order from west to east:

1	2	3	4
Fish	Snake	Bird	Snake

The solution to the puzzle is to match each pillar's face to the closest hint.

If you are struggling, stand at the lever and face south. Arrange the three pillars in the following order relative to that position.

Left	Center	Right
Bird	Snake/Bird	Bird

The central pillar's engraving will determine which door you will open. Snake will open the left gate, leading deeper into Skuldafn. Bird opens the right gate, leading to loot.

You can open both gates without any consequences.

After solving the puzzle, you will reach a small room with yet more draugr with a webbed-up doorway. As it suggests, you will need to kill Frostbite Spiders in the next room. You will encounter a second puzzle at the end.

❖ Solving The Second Skuldafn Puzzle

A collage showing the correct solution to a puzzle

Your progress will be halted in a large room with two floors and a raised drawbridge. The second puzzle is found here. Clear out the draugr before you start.

You can Whirlwind Sprint through the drawbridge at the right angle.

This second puzzle uses the same spinning pillars that the prior one did. The pillars are kept in alcoves marked by a stone engraving that matches one of the three faces of each pillar. Rotate the pillars to match the symbol of the alcove they are kept in. The West pillar is directly in front of the entrance.

North	West	South
Fish	Snake	Bird

Solving the puzzle will lower the drawbridge, allowing you to continue.

There will be more draugr as you continue through Skuldafn Temple. You will eventually reach a spiral staircase. Don't stand on it immediately. A trap will trigger, and it will be set alight. Climb up once the fire burns out to fight more draugr and look for a lever in a small room visible at the staircase. Pulling it will open the door to this room.

The next room is a long tunnel with an oil slick spread through it. Try to set the oil alight as soon as possible to avoid having it used against you. The third and final puzzle is located at the end.

❖ Solving the Skuldafn Dragon Claw Puzzle

A door showing the correct combination to solve a puzzle

Make sure to kill the high-level enemy at the puzzle. They are holding the Diamond Dragon Claw necessary to solve this wall. As always, like in Bleak Falls Barrow, check the palm of the claw for the correct pattern. In this case, from top to bottom, it is:

- Fox
- Butterfly
- Dragon

The final room contains a Word Wall for the Shout Storm Call. Follow the quest marker to leave Skuldafn Temple and reach the portal.

This is your only chance to claim the word for Storm Call, as you cannot return to Skuldafn after the quest.

Defeating Nahkriin

Emerging back into the open air, you fight four draugr at the foot of a staircase leading to a beam of light. Defeat them and ascend the stairs to fight Nahkriin, the Dragon Priest stopping you from reaching the portal. He will remove his staff from a mechanism, closing the portal and forcing you to fight.

You might notice that there are a pair of dragons in this area. They are not aggressive unless attacked.

Nahkriin is a powerful spellcaster. He will use long-ranged spells to damage you at a distance, having access to a wide range of spells. Making things worse, he will summon a Storm Atronach to help him in the fight. If you are struggling, there are a few strategies that make it easier to get past him.

Melee	Fighting him up close will be frustrating. He will float away from you while striking you with spells. He will not hesitate to

	use frost spells to deplete your stamina If you are able to withstand his attacks, you can stun him with power attacks as you whittle his health down. Taking breaks to heal and recover where possible would be wise. This strategy can be made easier by using Spellbreaker and chugging plenty of potions.
Magic	Fighting Nahkriin with magic is inviting one challenging duel. Thanks to his staff, depleting him of magicka using shock spells will not guarantee success. If you are a skilled destruction mage, the perk Impact will let you stun him repeatedly, making it harder for him to fight back.
Archery	Nahkriin will not willingly approach the player. You can take advantage of a small pillar by the stairs leading to his staff by sheltering behind it and shooting arrows until he dies. This is not a particularly honorable method of beating him but it does work.
Assassination	You will have a short window to sneak attack him if you are patient in your approach and are sufficiently skilled.
Ignoring him	If you run very quickly, you can jump into the portal before he can close it. Doing this will prevent you from collecting his mask and from completing the side activity at Labyrinthian.

When he is defeated, collect his mask and staff. Interact with the altar to open the portal and enter Sovngarde.

SOVNGARDE

Finding Your Way To The Hall Of Valor

A deep mist has spread across the realm, stopping lost souls from reaching the Hall of Valor where they are meant to reside. You need to find your way through the mist to reach the Hall Of Valor.

Along the way, you will encounter lost souls struggling to reach the Hall. One of them, a dead soldier, will approach you early in your exploration, saying that Alduin has been eating souls lost in the mist. You can offer to help this soldier by letting them follow you.

Keep your eyes peeled for familiar faces in the mist. If you need more motivation, Aludin's mist has ensnared the souls of old friends and enemies alike based on the quests you have completed.

Enemy commanders from the Civil War and even a member of the Companions can be found within.

Use Clear Skies to clear the mist away, revealing a path forward. Broadly speaking, simply following the path will get you to the Hall of Valor.

Proving Yourself To Tsun

You will find the way blocked by a massive, shirtless god named Tsun. On asking to enter the Hall, he asks why you have the right to enter. The answers you have available depend on the factions you have come to lead, though there are no wrong answers.

In true Nordic fashion, you must defeat him in battle to earn the right to cross the bridge. Tsun doesn't have any particularly unique mechanics, though he is still a challenging opponent. He will use a two-handed axe and some Dragon Shouts to fight you. Getting him down to just about half-health will make him yield and allow you entry.

Cross the Bone Bridge to reach the Hall of Valor.

Talking To The Heroes Of Sovngarde

If you've been paying attention to the Companions questline, you might recognize the man who greets you, Ysgramor. If there was ever any doubt that you are in the presence of champions, that should dispel them.

Though you will find many legendary figures spoken about in your journey, you need to find a trio of heroes to take the fight to Alduin. You will find them by the giant row of cooking hogs, indicated by the quest markers. You'll recognize them from Alduin's Bane, when you read the Elder Scroll.

Listen to them speak and devise a plan to defeat Alduin before the quest ends, starting the next one Dragonslayer.

DRAGONSLAYER

Starting The Fight With Alduin

To challenge Alduin, leave the Hall of Valor and cross the whalebone bridge. The Heroes will speak amongst themselves, trying to figure out how to dispell the mist. After some back-and-forth, they will decide to Shout together to combine their voices.

Use the Clear Skies Shout three times. The Heroes will follow your lead, managing to dispel the mist. Alduin will reform the mist twice and will descend to fight you on the third time.

Defeating Alduin

A man leaps to strike a black, menacing dragon in a grass field

Alduin doesn't have any new tricks though he isn't any easier to fight this time. You will want to use Dragonrend to ground him and render him vulnerable to your attacks. The Heroes will fight him as well. They will not permanently die and can be used as a distraction if things get dicey.

Bite	Alduin lunges forward to bite at you when you are too close.
Fire Breath	Alduin breathes fire breath, damaging you over time and inflicting lingering damage
Frost Breath	Alduin breathes frost breath, damaging you over time and sapping your stamina
Dragon Storm Call	Alduin calls a meteor storm. Firey rocks will rain from the sky damaging you and inflicting fire damage if you are caught in the impact.

This is, for all intents and purposes, the final boss fight of the game and you should hold nothing back. Use any potions you might need without reservation as you deplete his health.

His fire breath is particularly deadly. Like other dragons, you can stop him from using it with a well-timed shield bash or blasting him with Unrelenting Force. Just make sure to keep him under the effect of Dragonrend to force him on the defensive.

If you are struggling to defeat Alduin and you don't seem to be adequately prepared for the challenge, you can let the Heroes do all the fighting and land the final blow.

Alduin will be defeated when he runs out of health. He won't die until you deal the final blow. Screaming in denial, he explodes horrifically. With this, Sovngarde and Skyrim, are saved.

In the aftermath of his defeat, speak to Tsun to return to Skyrim. Before you leave, he will teach you the shout Call of Valor.

You will not be able to return to Sovngarde after you leave. It is a unique area that has many easter eggs for people who pay attention to the game's lore.

Epilogue

You return to the Throat of The World. Dragons surround you as they speak in their language before departing. Paarthurnax will speak to you about his brother's death for a time before leaving, declaring that he will convince other dragons to come around to his peaceful philosophy.

If you have finished the quest Paarthurnax, this scene will be skipped.

Odahviing will land next, saying that you can call on him anytime for help in battle. Using all three words of the Call Dragon Shout will summon him as an ally.

You may visit the Blades and the Graybeards to speak about your accomplishment as well.

SIDE QUESTS

MISSING IN ACTION

Starting Missing In Action

You start this quest by talking to Fralia Gray-Mane at her market stall in Whiterun after Olfrid Battle-Born mocks her missing son. She will be grateful for your sympathy and asks you to visit her house where you can learn more.

The quest can also be started by talking to the Battle-Borns after the argument or visiting House Gray-Mane directly.

At House Gray-Mane, her remaining son, Avulstein, will confront you. He is afraid that he will be taken next. He does not trust you until Fralia convinces him otherwise.

Reluctantly, he asks you to find evidence of his brother's survival. Based on how the Battle-Borns have been treating the family and because of their grudge, he suggests searching House Battle-Born.

Finding The Evidence

While Avulstein suggests a practical and fairly straightforward approach, you can find alternative methods of discovering where Thorald is.

❖ Stealing The Evidence

You are looking for a missive inside House Battle-Born. It is on the first floor, in the west wing of the house, and locked behind a leveled lock. If you are a skilled lockpick, this would not be difficult.

There is no key for this office, meaning that you cannot pickpocket a Battle-Born or find the key elsewhere. You have to pick the lock.

This is further complicated by the house being off-limits to strangers. You will have to sneak through the house unless you have befriended the family already. There are three methods to gaining their friendship.

Help Lars	Their youngest son, Lars Battle-Born, is regularly bullied by Braith. You can learn about this from him and help him put a stop to the bullying, earning his friendship.
Give the correct greeting to Idolaf	Idolaf Battle-Born will ask to know which side of the feud you are on when you first speak to him. If you answered "Battle-Born", you will become

	friends.
Sell crops to Alfhild Battle-Born	You can find the Battle-Born Farm to the southeast of Whiterun, by the city walls. Selling any crops to her will make you a friend.

You will be able to walk through the house after doing any of the above, making your heist much easier.

❖ Blackmailing Jon Battle-Born

Jon's Romeo-and-Juliet-esque romance with Olfina Gray-Mane is a small detail that is easy to miss when traveling through the city. You can take advantage of it by pickpocketing a letter written by Olfina from him and confronting him with the evidence.

He will panic and bring you the missive at the Statue of Talos at night, getting you the evidence.

❖ Convincing Idolaf Battle-Born

You can bring up Thorald's disappearance with Idolaf by talking to him about it. He refuses to discuss it with you, though you can convince him otherwise.

This method requires your Speech to be at least level 75 to succeed.

If you succeed, Idolaf will give you a copy of the missive and tell you that Thorald is at Northwatch Keep, a place fearsome place where people disappear. He believes that Thorald is dead.

Rescuing Thorald

An adventurer stands off against a soldier outside a snowy castle.

Return to House Gray-Mane to share your findings with Avulstein. He will spring into action and prepare to rescue his brother. You can agree to follow him or ask him to wait at home while you rescue Thorald.

You may wish to allow Avulstein to come with you. He is an essential character and cannot be killed, making him very useful in clearing out Northwatch Keep.

Regardless of your decision, you will go to Northwatch Keep to mount your rescue operation. It will not be easy. The keep is manned by Thalmor agents that cannot be convinced to let you in.

Northwatch Keep is on the northeastern most tip of Haafingar Hold, far beyond the mountains. You can reach it by fast-traveling to the Thalmor Embassy from Diplomatic Immunity or Wolfskull Cave from The Man Who Cried Wolf.

The Thalmor guards are equipped with either Elven or Glass armor and weapons and are difficult opponents to fight. A frontal assault would be challenging at best and may result in Avulstein's friends dying. You will need to cross the courtyard and enter the keep itself to find Thorald imprisoned at the end of it.

You could try sneaking through the fort by going to a side entrance on the north side and picking the lock. It will give you an alternative route into the fort that has fewer enemies.

Completing Missing In Action

Once you reach Thorald, talk to him for him to be freed before escorting him out of the fort to safety. Once outside, he will reason that he cannot return home as the Thalmor are looking for him. He gives you a message for his mother and leaves to join the Stormcloaks.

Return to Whiterun to give Fralia the news. She will be saddened by his decision, though she is happy that her son is still alive. She will give you an enchanted steel weapon as a reward for saving her son, ending the quest.

THE BLESSINGS OF NATURE SIDE

Starting The Blessings Of Nature

You are likely to pass Danica Pure-Spring on your many trips to the Cloud District or Jorrvaskr at Whiterun. The priestess of Kynareth loudly laments the Gildergreen's sad state each time you pass by when she is sitting by the tree.

She will be glad for a listening ear and explains that she has a plan to restore the Gildergreen. She needs a unique dagger called Nettlebane to cut a way to the Eldergleam and extract some sap to rejuvenate the Gildergreen.

Getting Nettlebane

A man leaps through the air, trailing fire, as he strikes down at a mutated and misshapen woman.

As much as Danica wants to handle the issue herself, Nettlebane is being held by hagravens at Orphan Rock. She does not have the martial skill needed to defeat the hagravens and asks you to take it on her behalf.

Orphan Rock is located at the foot of the Throat of the World. You can get there by fast traveling to Helgen and following the road southeast. It is a short distance away.

You will need to dispatch several magic-wielding hags when you arrive. Clamber up Orphan Rock, past the Forsworn-themed tents, to find a single hagraven at the top. She is a challenging foe, especially at lower levels. You will need to defeat her to get Nettlebane. With the weapon in hand, return to Danica in Whiterun.

Danica will be disgusted by the dagger and refuse to touch it. She asks you to complete the final step and head to the Eldergleam Sanctuary to get the sap.

A nearby worshiper, Maurice Jondrelle, will ask to join you as he has always wanted to see the Eldergleam. You can choose if he will join you.

Maurice will not take up a follower spot, though he is a weak fighter that should be kept away from enemies. You should go directly to the Sanctuary if you are allowing him to tag along.

The Eldergleam Sanctuary

The Eldergleam Sanctuary is east of Whiterun, located in the hot springs to the south of Windhelm. The Sanctuary itself does not have any enemies within. You only need to follow the path to the roots and attack them with Nettlebane. The roots will part as soon as they are hit, giving you a path to the Eldergleam.

If you brought Maurice along, he will be horrified and demand you to stop. If you explain the situation to him, he will propose an alternative solution. The quest will have two possible outcomes based on what you choose.

Take the sap	Attacking the Eldergleam with Nettlebane will get you the sap, though two spriggans will appear and attack you. If Maurice was brought along, he will turn hostile as well. It is possible to end the quest without killing him by sheathing your weapon after the spriggans are defeated.
Allowing Maurice to pray	If you allow Maurice to pray at the Eldergleam, you will be rewarded with a sapling and will not have to fight anything. Though disappointed, Danica will accept the sapling.

In both instances, Danica will be pleased with the outcome and work on restoring the Gildergreen. After several days, the tree will be in full bloom. She will become available as a Master-level Restoration trainer, and you will be able to keep Nettlebane as a weapon.

IN MY TIME OF NEED

Prerequisites To Complete

Prior to starting In My Time Of Need, you will need to complete Dragon Rising. During this quest, you will battle against your first dragon and absorb a Dragon Soul. To start the Dragon Rising quest, all you need to do is speak with the Jarl of Whiterun.

Once Dragon Rising is complete, you can immediately begin In My Time Of Need.

Speaking With Alik'r Warriors

Your first task is to find and speak with Alik'r warriors just inside the gates of Whiterun.

The guards won't let you further into the actual city at this point.

Talking to them, you'll learn they're looking for a Redguard woman who they claim is a refugee from Hammerfell, though they won't tell you why they're looking for her.

You can find these warriors in the Rorikstead Inn after this encounter.

Finding Saadia

The Redguard woman in question is Saadia, who resides in The Bannered Mare tavern of Whiterun - the very city the Alik'r were barred from entering. Talking to her about a wanted Redguard woman will make her nervous, so she'll ask to speak to you in private in her upper-floor room.

You can follow her there or report her location to the Alik'r Warriors in the Rorikstead Inn.

If you want the quest done ASAP, the latter option will end it quickly and give you a 500 gold reward.

Speaking With Saadia

If you want to play out the rest of the story, follow Saadia and she'll tell you her real identity is Iman, a noblewoman of Hammerfell's House Suda. Allegedly, she spoke out against the Aldmeri Dominion and was forced to flee from the assassins they sent after her.

The leader of the Warriors after her, Kematu, has claimed she betrayed her people to the Thalmor, which she denies. If Kematu is killed, Saadia says the other assassins would scatter without leadership.

She points you to the Dragonsreach Dungeon where a rowdy Alik'r Warrior was arrested for trying to enter Whiterun.

Talking To The Prisoner

This Alik'r Prisoner is in one of the dungeon cells of Dragonsreach Dungeon. You can persuade the prisoner to reveal the location of Kematu by paying his bail of 100 gold. After this, he'll say Kematu is in Swindler's Den.

If you have cleared Swindler's Den prior to meeting Saadia, talking to the prisoner is skipped entirely, and you will state that you know Kematu's location already.

Traveling To Swindler's Den

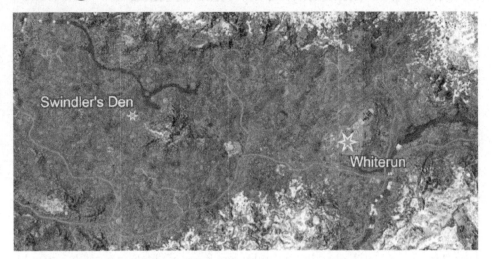

This classic Skyrim dungeon is west of Whiterun at the bottom of a rock hill; you can see this location on the map above.

There are many Bandits to fight before reaching Kematu, as well as a secret shortcut that cuts down on the amount of bandit-killing you have to do.

This shortcut is in the first room, behind the chair, table, and bedroll, and actually functions as the exit to the dungeon after defeating all the Alik'r Warriors.

Skyrim is merciful enough to have shortcuts that prevent excessive backtracking in the game's dungeons, and Swindler's Den is thankfully no exception.

You'll see a large boulder with moss growing above it; you can get up to the entrance with some skilled jumps, skipping several fights along the way and immediately entering the room with Kematu.

You can also find the book Thief in Swindler's den, which details the continuing adventures of Eslaf Erol and will level up the Pickpocket skill. The book can be found next to a bedroll in the most eastern room of the Den.

Other than this, Swindler's Den has little in the way of valuable treasures, other than some Ingots and gold pieces scattered throughout.

Kematu's Offer

Instead of attacking you on sight, you'll have a chance to side with Kematu and hear his version of events: that Saadia is a traitor to her people and caused the fall of the city Taneth to the Aldmeri Dominion.

You can either agree to lead Saadia to Kematu outside of Whiterun or kill Kematu then and there.

Either way, your reward will be 500 gold, so it's solely up to who you believe more.

Rewards For Completing In My Time Of Need

❖ Killing Kematu

If you killed Kematu, you can return to Saadia and ensure her safety, upon which she'll reward the Dragonborn with 500 gold.

❖ Helping Kematu

If you agreed to help the Alik'r, you will tell Saadia that she is in danger and that there's a horse waiting for her outside Whiterun to escape on. Kematu will be waiting for both of you at the stables, enraging Saadia before she is paralyzed by Kematu. He will award you 500 gold after this, completing the quest.

You can earn more gold by killing Kematu at the stables before getting the reward from him. Saadia will be angered about being deceived, but will reward you 500 gold for killing Kematu.

After this, players can loot additional gold from Kematu's body, a possible 250 gold.

❖ Doubling The Reward

There is a method of obtaining 500 gold from each of them, but it involves killing Kematu and having Saadia permanently in a hostile state.

You can kill Kematu after collecting your reward from him, then use Calm on Saadia as she will become hostile after recovering from paralysis.

Calm allows you to collect the extra reward of 500 gold, but when it wears off she will be permanently hostile. Killing her in this state is fortunately not a crime, though it is more senseless violence. All par for the course in Skyrim.

NO ONE ESCAPES CIDHNA MINE

Prerequisites To Complete

Prior to starting No One Escapes Cidhna Mine, you will need to complete The Forsworn Conspiracy. This quest will trigger once you enter Markarth for the first time, and involves solving the mystery surrounding Weylin and the Forsworn.

At the end of Forsworn Conspiracy, the guards of Markarth will attempt to arrest you. If you submit to the guards, this quest will begin. If you run from the guards, the quest will not start. This means that you need to go to jail for this quest to begin. Once arrested, you will arrive in Cidhna Mine and the quest will begin.

Talking With Urzoga And Uraccen

Your first task is to speak with Urzoga gra-Shugurz. During this conversation, he will tell you that you need to mine yourself to freedom.

End the conversation and head through the gate until you come to Uraccen. Uraccen is a miner who provides information on Madanach, the leader of the Forsworn. Talk with this NPC, from which you learn that to get to Madanach, you need to get past Borkul the Beast.

Borkul The Beast

Borkul the Beast will be guarding Madanach's cell. There are a few different ways to get past him. To enter Madanach's cell, you will need to give Borkul a shiv. When speaking with him, you can either present the shiv, or tell him you don't have one.

Alternatively, you can persuade, bribe, intimidate, or brawl with Borkul. If successful, he will let you pass.

If you attack Borkul outside of the brawl, every prisoner, including Madanach, will attack you. When this happens, you will be locked into killing Madanach later in the quest.

❖ Finding A Shiv

To find a shiv, you will need to speak with Grisvar the Unlucky. He will ask that you bring him a bottle of skooma in return. Skooma can be acquired from Duach, who is also found in the mines. Once speaking with him, you will either need to persuade, brawl, or pickpocket the skooma.

With the skooma in hand, head back to Grisvar, trade for the shiv, and then return to Borkul.

Confronting Madanach

After getting past Borkul, move forward and speak with Madanach. There are two possible options here; listen and find answers, or kill Madanach.

❖ Listening To Madanach

If you choose to spare Madanach, he will tell you to speak with Braig and listen to his story. Upon listening, you will start to see a different side of Madanach. To prove your loyalty to this character, he will ask you to kill Grisvar the Unlucky (this NPC provided the shiv earlier).

Madanach will give you a new shiv if you gave yours to Borkul, and send you off to kill Grisvar. Once he is dead, speak with Madanach; he will make you an honorary member of the Forsworn. Now, leave the prison through the tunnel with the other prisoners.

❖ Killing Madanach

If you choose to kill Madanach, you can either tell him you are going to, or surprise attack him. Once dead, search his body for a note and key. Use this key to unlock the gate and leave the Cidhna Mine through a tunnel heading into the Marlarth Ruins.

Leaving Cidhna Mine

The exit tunnel will contain a few enemies, including frostbite spiders. Once in the Markarth Ruins, you will also encounter Dwarven Spheres.

If you have escaped after killing Madanach, you will have your equipment returned by Thonar, and be rewarded with the Silver-Blood Family Ring.

However, if you escape with Madanach, your equipment will be returned by Kaie, a member of the Forsworn. She will also reward you with the Armor of the Old Gods. The quest will continue until Madanach is confronted by Thonar, and the two begin to fight.

This fight will kill Thonar, and turn several guards and citizens hostile towards the Forsworn. You can participate in this battle, or wait until midnight near the exit tunnel until the fighting dies down.

Rewards

Depending on the route you took, you will either receive the Silver-Blood Family Ring or the Armor of the Old Gods. Let's take a look at each of these rewards below.

❖ Silver-Blood Family Ring

This family heirloom will provide the Fortify Smithing enchantment, which improves your smithing by 15 percent.

❖ Armor Of The Old Gods

This leather armor set will provide the following enchantments.

Armor Piece	Enchantment
Chest Piece Of The Old Gods	Destruction spells will cost 15% less to cast.
Boots Of The Old Gods	Sneak 20% better.
Gauntlets Of The Old Gods	Bow attacks will do 20% more damage.
Helmet Of The Old Gods	Magicka will increase by 30 points.

Is It Possible To Get Both Rewards?

It is possible to get both rewards, but it can be a bit tricky. To do so, do not kill Madanach while in the prison. Once you leave through the tunnel, accept the Armor of the Old Gods, and then start a conversation with Madanach. This will stall him as others run by.

Once everyone is out, kill Madanach and leave the area. At this point, you should not loot his body. Forsworn members will begin to attack you. Defeat them and then return to the body of Madanach to loot the note and key.

Do not kill Forsworn members while in the Markarth Ruins. This will result in not receiving both rewards.

Read the note and then exit the ruins. At this point, Thonar will appear; speak with him to receive the Silver-Blood Family Ring.

THE FORSWORN CONSPIRACY

Arriving In Markarth

When you arrive in Markarth and head to the market, you will witness a person named Weylin shout, "The Reach belongs to the Forsworn!", before attacking a woman named Margret.

During this scene, you will have a few different options. First, you can attack Weylin before he can make his move. This will save Margret, and she will reward you with a silver emerald necklace.

Another option during this scene is to sit back and do nothing. If you choose to do nothing, you will witness the murder of Margret. After she is murdered, guards will chase Weylin and kill him as well.

You can loot Margret's body for a key to her room at the inn. This can be used later, but if you don't pick it up, you can obtain another one from the barkeep at the Silver-Blood Inn.

Regardless of what choice you pick, Weylin will die, and you can continue on through the rest of the quest.

Meeting Eltrys

After watching the murder (or attempted murder) of Margret, a Breton named Eltrys will approach you. He will hand you a note, asking you to meet him at the Shrine of Talos.

During this interaction, you can ask him how to get there. His answer is a bit vague and he will pretend like he doesn't know, but he will mention that it might be under the Temple of Dibella.

Locating The Shrine Of Talos

To get to the Shrine of Talos under the Temple of Dibella, head to the staircase in the northern corner of the market. Once up these stairs, head left across a bridge, and then take another left until you come to a door.

Proceed through the door, and you will find Eltrys waiting for you. Here, he will tell you about the Forsworn, and wants your help to find out why Weylin attacked Margret. From here, you will need to gather information from the following places:

- Silver-Blood Inn
  - Information on Margret.
- The Warrens
  - Information on Weylin.

Finding Information About Margret

To learn about why Weylin attacked Margret, you will need to find information about her. This can be done by going to her room at the Silver-Blood Inn.

❖ What To Do If Margret Is Dead

If Margret is dead, you will need to speak with the barkeep. If you already have the key to her room from looting her corpse, you can head straight to her room. Alternatively, you can also just pick the lock to her room.

Inside her room, you can find her journal, which mentions a person named Thonar Silver-Blood.

❖ What To Do If Margret Is Alive

If Margret is alive, you can find her in the inn next to the fireplace. Regardless of if she is alive or not, you will gain the same information about Thonar Silver-Blood.

Finding Information About Weylin

The Warrens is an area in Markarth where the sick and poor live. Once you make it here, speak with Garvey to gain the key to Weylin's room. Alternatively, you can pick the lock to his door. To get to Weylin's room, head to the last door on the right side.

Inside, you will find a chest containing Weylin's Note. Upon reading it, you will learn of a person called "N". With this information, exit the Warrens until you are approached by Dryston. After warning you to stop 'digging around', Dryston will attack you.

If you attack him with any type of weapon, you will receive a bounty. Additionally, any nearby guards all start to attack you. Be sure to attack with just your fists!

Once Dryston is defeated, he will tell you that he was sent by Nepos the Nose.

After learning information on Margret and Weylin, head back to Eltrys. He will give you a bit of gold as a reward, and then you can continue on learning about Thonar and Nepos. If you ignore Eltrys and go straight to Thonar or Nepos, you will not receive this reward.

Thonar Silver-Blood

Now it's time to learn about Thonar, who can be found in the Treasury House. Here, you can speak with Rhiada to gain access to Thonar's room. Similar to the rooms of Margret and Weylin, you can also pick the lock to Thonar's room. To get there, take a left and proceed through the door directly in front of you.

Once you have access to his room, you can find him sitting at the table. Speak with him until you hear noise coming from outside.

At this point, Thonar will run outside, where his wife is getting murdered by Nana Ildene and Donnel. It is safe to kill both of them; they are undercover Forsworn agents. After dealing with Nana and Donnel, Thonar will tell you about Madanach, the leader of the Forsworn.

The death of Thonar's wife is scripted; once you speak with him, she is destined to die. To avoid this, you can pickpocket his journal. This is tricky, but it will prevent his wife from dying since you don't have to speak with him.

Nepos The Nose

Lastly, it's time to gather information on Nepos the Nose. Head to his house, located on the third terrace next to Vlindrel Hall. Once here, speak with his housekeeper, Uaile. At first, she will keep you from speaking to Nepos, but he will eventually interfere and welcome you in.

During your conversation, Nepos will admit that he is Forsworn, and tell you about Madanach. Additionally, he will tell you that Uaile is an agent. At this moment, everyone in the house will become hostile. Defeat them all, and then read Nepos' journal.

Prior to leaving, be sure to search his home for loot. You can find up to five silver ingots, as well as several residents carrying expensive items.

To avoid fighting with Nepos and company, you can pickpocket his journal. If you chose to do this, you will need to move quickly and leave immediately afterward.

Returning To Eltrys

With information on Thonar and Nepos, head back to the Shrine of Talos. Once you make it here, you will see guards standing over the dead body of Eltrys. After a brief conversation, the guards will try to arrest you. If you submit to arrest, you will begin the No One Escapes Cidhna Mine quest.

Choosing to resist arrest will result in several waves of guards attacking you. With all the guards defeated, you can leave the Shrine of Talos.

Because you resisted arrest, there will be a bounty on your head. This is a unique bounty that will not be affected by bribery or intimidation. While in Markarth, any guard that you see will attempt to arrest you on sight. You can continue to avoid being arrested, but this bounty will not disappear.

To end the endless arrest on sight order, you will need to submit and go to jail, which begins No One Escapes Cidhna Mine.

THE HEART OF DIBELLA

Starting The Heart Of Dibella

You will need to infiltrate the Inner Sanctum in Markarth's Temple of Dibella to start the quest, which is not something you would do under normal circumstances. As soon as you set foot within the temple, you would be told that the priestesses are occupied with a ritual and that they are not to be disturbed.

Accepting a quest from Degaine, unlocked by talking about the beggar with Kleppr at the Silver-Blood Inn will have you conducting this infiltration. Degaine will ask you to steal a Statue of Dibella from the Inner Sanctum, setting you on the path to starting this quest.

You will need to pick the lock to the Inner Sanctum at the door opposite the entrance and walk in. Senna, the head priestess, will stop you if you are detected. She is furious that you have intruded on their ritual and gives you a choice; help the Temple or get executed. You must agree to help the Temple to continue the quest.

Resisting the priestesses will lead to a fight. They are weak fighters that will not pose a threat and must be killed. You cannot continue the quest if they are dead.

Finding The Sybil

Senna explains that the Temple needs a new Sybil. Their ritual has found a new one in Karthwasten, a small town to the northeast. You will need to bring her back so that she can be inducted into the order.

You will need to ask around town to find clues about the girl. Any of the residents at the town will tell you that the only person matching your description was recently kidnaped by the Forsworn. They direct you to her father, Enmon, where you learn more.

There are two options for gaining Enmon's help. You could introduce yourself as a concerned bystander or tell him that his daughter is the new Sybil. Both options will have him assist you though he will ask to accompany you in the latter.

Enmon can die if he accompanies you. It may be best to ask him to remain at

Karthwasten instead of allowing him to follow you.

Rescuing Fjotra

The Forsworn have brought her to Broken Tower Redoubt, where she is has been imprisoned. You will need to fight her captors throughout the dungeon before you can reach her.

Inside the Redoubt, you must head up the stairs to the second floor and circle westward to reach the route up to Fjotra. You will need to defeat a Briarheart to free her. With the Sybil rescued, you need to bring her to the Temple of Dibella.

Though not necessary, you can bring Fjotra to her father to let them say goodbye to each other.

At the Temple, report your success to Senna. She will forgive your transgressions and will allow you to receive a blessing from Dibella, ending the quest.

This blessing, Agent of Dibella, allows you to deal more damage to the opposite sex.

THE MAN WHO CRIED WOLF

Starting The Man Who Cried Wolf

You start this quest by speaking to Falk Firebeard after your first visit to the Blue Palace in Solitude. You will see Varnius Junius from Dragon Bridge ask the court to investigate Wolfskull Cave. While Jarl Elisif is initially convinced, her court dissuades her from sending an Imperial Legion. She instead sends more guard patrols.

You can start the quest at any time after you witness this exchange.

Falk will allow you to investigate when you volunteer and promises a small reward for doing so. You will need to depart for Wolfskull Cave, found in the mountains to the north of Dragon Bridge.

Entering Wolfskull Cave

You will be attacked by a pair of skeletons when you pass by the entrance. They are the first of many undead that you will fight throughout the quest. The rest of the cave follows a similar theme, with necromancers and undead forming the bulk of your enemies.

Start fighting your way through the necromancers and their undead allies until you pass through a small ruin and see a deep sinkhole. You will have to jump into the hole to continue.

You will not be able to leave the cave until you complete the quest.

There will be no opposition between you and Wolfskull Ruins, the next area.

Clearing Wolfskull Ruins

It is apparent from the first few seconds inside the cave that this is no ordinary necromantic ritual. Between the huge concentration of necromancers and the massive, purple ball of light in the center, whatever magic occurring here is more potent than anything you have seen before.

The ritual is summoning Potema Septim, a ruler of Solitude that had infamously practiced necromancy.

You will have to put a stop to their ritual. To do so, take the cave to the left. It will allow you to descend to the massive castle-like ruins at the center of the chamber with only a single necromancer in your way.

Be mindful of the Burning Lamp traps suspended on the ceiling. They can be easy to miss in the chaos of the ruins.

From here, head toward the ruins and focus on climbing up to the top of the tower. You will need to cross the stone bridge to go up the stairs leading to the stone tower

Try to focus on the necromancers, as they will raise fallen enemies to harass you if left alone. They are physically weak enough to be slain without much of a fight.

At the very top of the tower, you have to defeat the Ritual Master. She will send a pair of weak necromancers to slow you down. Once she is slain, the ritual will end, and your job here will be done.

Finishing The Man Who Cried Wolf

To finish the quest, pull the lever on the left of the raised drawbridge. It will allow you to reach the next tower. Go down the stairs to find a shortcut back to Wolfskull Cave. All you need to do now is return to Falk Firebeard for your reward.

He will be surprised and relieved that you managed to stop the Potema from being summoned, giving you your payment.

THE WOLF QUEEN AWAKENED

Starting The Wolf Queen Awakened

You will be approached by a courier sent by Falk Firebeard, asking you to return to the Blue Palace for help. There are new signs that Potema is returning and, given your prior encounter, your services will be especially valuable. Go to the Blue Palace in Solitude and start the quest by talking to him.

This quest needs you to complete The Man Who Cried Wolf and to gain one level before the courier is sent.

He asks you to speak to Styrr in the Hall of the Dead. It is just down the road from the Blue Palace. Styrr explains that one of Potema's undead minions had emerged from the Solitude Catacombs in the Temple of the Divines. You have to clear them out and stop Potema.

Clearing The Catacombs

The Temple of the Divines is close to Castle Dour. Enter the building, take the hallway on the west, descend a staircase, and use a key Styrr gave you to unlock a metal gate. The destroyed wall is right in front of you. Enter it to reach the catacombs.

As soon as you step foot inside, Potema gives a greeting. She opens the way into the catacombs and is eager to "reward" you for preventing the necromancers from enthralling her soul in Wolfskull Cave.

You will fight a combination of draugr and vampires, a combination that befits a queen infamous for necromancy. These enemies are vulnerable to silver weapons, Flame damage, and other weapons that damage the undead such as Dawnbreaker.

Dust and cobwebs aside, the catacombs resemble Solitude's many opulent buildings only filled with undead instead of nobles. It is a linear dungeon that does not have any unusual quirks to be aware of.

There is a hallway blocked by a rotating door. Simply time your movement through it to get through. The next area, Potema's Refuge is right after it.

Solving The Rotating Doors At Potema's Refuge

Potema's Refuge is a return to the ancient Nordic ruins that you have seen before. Cleanse it of undead until you reach a trio of levers and a matching set of rotating doors.

Each door is controlled by a corresponding lever, in the same order that they are encountered. For example, the lever furthest from the doors controls the outermost door. Each lever can be set to rotate clockwise or counterclockwise and will stop the door when raised to neutral. Stop the doors when a way through is established.

The final obstacle is a Blooded Vampire who will attempt to kill you to earn Potema's favor. They are assisted by a powerful draugr.

The door leading out of this chamber has a master-level lock. The Blooded Vampire carries its key.

The next door leads you to the final area, Potema's Sanctum where the Wolf Queen awaits.

Defeating Potema At Her Sanctum

There are only a handful more undead enemies left. Your path through will be blocked by Potema who raises a trio of bodies in a room full of corpses. You will have to defeat them to progress.

The corpse room's tight confines can be difficult for archers and mages to work around. You can run out of the room before they return to their feet or use Unrelenting Force to make some space.

Potema is all that remains. She floats above a large chamber as a ball of magic. The Wolf Queen will unleash her Inner Circle against you, sending several waves of draugr into the chamber to defeat you.

Potema will not be idle. She fires beams of lightning at you throughout the fight. These beams are incredibly damaging and will drain your Magicka. You will need to run out of their path to avoid being hit.

A man strikes down a ghostly woman with a throne and skull in the background.

Once you have ended her Inner Circle's unlife, she will flee the chamber and face you as a ghost. At this point, she poses little threat to you and can be dispatched easily.

Make sure to take her skull and help yourself to the contents of a large chest next to her throne.

With the threat ended, you can exit the catacombs and emerge outside Solitude, on a cliff.

Completing The Wolf Queen Awakened

Falk is waiting for you at the Blue Palace, hoping to hear the good news. More importantly, he has your payment and a bonus, the Shield of Solitude.

The Shield of Solitude increases your magic resistance by 25 percent and will block 35 percent more incoming damage. It also sports the city's crest.

You will also need to return to Styrr and provide him with Potema's Skull so that he can purify it. With both tasks done, the day is once again saved and the quest is completed.

LIGHTS OUT

Starting Lights Out!

You will need to speak to Jaree-Ra in Solitude to start the quest. The Argonian spends his days waiting by Angeline's Aromatics, trying to find someone to help him with a "business venture".

If you take the bait, he explains that a fully laden cargo ship, the Icehammer, is sailing down into Solitude. You will need to ensure that the Solitude Lighthouse has its fire extinguished to make it runs aground.

The Solitude Lighthouse is north of the city, accessed by following a road heading north from the stables and past the docks. A nearby fast-travel point would be Brinewater Grotto, visited during the Thieves Guild quest Scoundrel's Folly.

At the Lighthouse, climb the stairs to the top and interact with the fire to extinguish it.

Claiming The Loot

Return to Jaree-Ra at the Solitude Docks. He asks you to meet with his sister, Deeja, at the Wreck of the Icerunner to claim your share of the loot.

You find the Icehammer crawling with Blackblood Marauder bandits and devoid of loot. Deeja is waiting deep within the ship. She gleefully tells you that she needs to handle one last loose end: you. You will have to defeat her and search her body for a Note from Jaree-Ra.

The Icehammer can be challenging to navigate. To reach Deeja, take the stairs opposite the door and walk south to the last door on the left, with a dead body inside. Find a staircase in this room, heading down. Deeja is at the opposite end of the lowest deck.

You will have to raid Broken Oar Grotto to get your share of the loot and take your revenge.

Clearing Broken Oar Grotto

The last of the Blackblood Marauders and their scaly leader have turned a trapped ship into their home at the grotto. You should stay on the right side of the grotto, encountering token resistance from the sentries.

Jaree-Ra is at the very end, on the ship with a platoon's worth of bandits. The ship's layout makes stealth challenging and a brawl highly likely. If a head-on assault looks risky, try to pick off enemies from a distance until you are discovered.

Make sure to loot Jaree-Ra for a key to a locked chest with the treasure and to end the quest. With your employer dead, your reward will have to be looted from the grotto.

LAID TO REST

How To Start Laid To Rest In Skyrim

When you arrive in Morthal for the first time, you may notice a the remains of a house that burned down. Head to the Moorside Inn and ask its innkeeper, Jonna, about it. She'll inform you that a man named Hroggar owned the house and that his wife and child died in the fire.

People suspect Hroggar might have caused the fire, as the very next day, he hooked up with a woman named Alva. Jonna will point you to the jarl, Igrod Ravencrone (or Soril The Builder if you've helped the Stormcloaks secure the city) at Highmoon Hall if you're interested in investigating.

Talk to the jarl and they'll wish to know what really happened, thus tasking you with finding out the truth of what really happened at Hroggar's house.

Investigating The House And Morthal

After you talk to the jarl, head over to the burned down house to investigate. Once there you will find the ghost of Hroggar's daughter, Helgi, in the corner of the house. You can ask her what happened, but she doesn't tell you much. Helgi will then ask you to play hide and seek with her at nightfall as her friend will be there as well — definitely not creepy at all.

Now, head back to the inn and ask Jonna about the ghost, and she will again point you to the jarl. Talk to the jarl again and ask about the ghost to find out that Helgi will be at the cemetery at night. Be sure to wait until nightfall if it isn't that time already.

Once nightfall approaches, head to the cemetery, which is above and to the right of the burned house. There, you will find a child's coffin where a woman named Laelette will attack you. It turns out this Laelette is a vampire. Once the vampire is dead, activate the coffin to hear about what happened from Helgi.

Helgi will reveal to you that Laelette was tasked with burning down the house in order to kill Helgi and her mother. Laelette tried to change Helgi into a vampire at the last moment and succeeded, but waited too long, as the child died in the fire anyway.

After this, a man named Thonnir, who turns out to be Laelette's husband, will stumble upon the scene and be obviously devastated over what happened to his wife onceyou tell him. He'll tell you Laelette disappeared at some point and that she started to talk to Alva just before her disappearance. He'll reveal that he actually talked to Alva the night she disappeared.

❖ Investigating Alva's House

Head over to Alva's house after speaking to Thonnir to investigate it. You may have to pick the lock to get in so make sure no guards are around when doing so. When you head inside, Hroggar will be there and attack you on sight, meaning you'll have to kill him.

Head into the basement and you will find Alva's coffin and possibly Alva herself inside of it, depending on the time of day you enter the house. In the coffin is Alva's journal, which reveals she, too, is a vampire and serves a master vampire named Movarth. The journal reveals how the two plan on turning Morthal into a blood farm to have a constant supply of fresh blood on hand.

Return to the jarl with this revelation and they will reward you with some gold that varies in amount based on your level. Now the jarl will ask you to deal with Movarth, which serves as the second part of the Laid to Rest quest in Skyrim.

Movarth's Lair

Movarth's Lair is located to the northeast of Morthal. Inside, he dwells with vampire thralls and other vampires. Before you head there though, you'll notice that Thonnir has gathered an angry mob outside of Highmoon Hall to help you out. Here, you can tell them to stay, since they will most likely die if they come. You can also allow just Thonnir to come. Either way, head over to the lair to take on Movarth.

Head inside the cave, where you will encounter some frostbite spiders. Keep pushing forward until you run into a dining hall area where Movarth is seated. Alva may be found in one of the back rooms (the one with bedrolls in it) but will be killed by the other vampires if you didn't encounter her earlier.

Movarth is a skilled conjurer and he can resurrect dead allies nearby to increase his numbers. Be sure you have plenty of healing potions on hand as well as some resist magic potions and drain magicka poisons, as that is his best weapon. Movarth can turn invisible, so be sure you have Laid to Rest as your active quest, as the quest marker tells you where he's at. You can also sneak to a hidden spot and shoot him with an arrow if you have a high enough sneak rating. Keep hitting Movarth hard until you finally kill him and save Morthal.

Completing Laid to Rest

Now it's time to head back to the jarl to deliver the good news. On the way out of the cave, you will encounter Helgi again and she'll thank you for what you did. She will then say she is ready to sleep before disappearing, having properly been laid to rest in peace.

Upon reporting back to the jarl, you will receive more gold, which again varies based on your level. You'll now be one step closer to becoming the thane of Morthal Should you own the Hearthfire DLC or any of the special editions of Skyrim, you will also be able to purchase Windstead Manor from the jarl's steward for 5,000 gold.

KYNE'S SACRED TRIALS

Starting Kyne's Sacred Trials

This quest starts in the mountains to the west of Riften and south of Ivarstead at Froki's Shack. Its owner, Froki Whetted-Blade, lives with his grandson far away from the rest of society.

Froki's home is in a remote location and is difficult to reach.

He strongly encourages you to complete the Sacred Trials of Kyne. This is a traditional challenge for hunters and is meant to test young hunters before they are considered ready. To sweeten the deal, he will provide you with a unique amulet for completing the trials.

Defeating The Pests

The old hunter anoints you, providing the preparation necessary to fight your first set of targets. They are weak enemies that barely pose a threat.

These enemies are merely ghostly versions of their usual selves and can be defeated using standard tactics.

Name	Location
Guardian Wolf	Found near Pinewatch, to the north of Falkreath.
Guardian Skeever	It is hiding inside the Windward Ruins to the south of Dawnstar.
Guardian Mudcrab	It is close to Gjukar's Monument.

Return to Froki's Shack once you have defeated them to continue the Trials.

Fighting The Predators

Froki wastes no time and sets you up against a trio of harder enemies, intended to test your prowess in combat.

Name	Location
Guardian Bear	Found near Knifepoint Ridge, near Falkreath, at a small glade.
Guardian Skeever	It hides in the glaciers southeast of Winterhold.
Guardian Mudcrab	The Mammoth waits by an area littered with mammoth bones to the north of Whiterun.

Return to Froki's Shack for the final step.

If you struggled with this set of enemies, consider getting stronger before starting the next part.

Vanquishing The Troll Champion

This final enemy is the most difficult part of the trials. You will fight the Guardian Troll at a small cave called Graywinter Watch, to the east of Whiterun. The battle against it is complicated by the two normal trolls waiting in the cave.

Trolls are weak to Flame spells and have their health regeneration halted when alight. Using spells such as Flames and Fire Breath will be a good method of keeping them at bay.

While not required, you can find Froki's Bow inside the cave on a barrel. It is a longbow that deals ten points of stamina damage but is otherwise unremarkable.

Completing Kyne's Sacred Trials

Return to Froki for a final time to complete the quest. There is no monetary reward though you will receive Kyne's Token, an amulet suited for hunters. It reduced damage from animals by 10% and increases bow damage by 5%.

THE BOOK OF LOVE

Starting The Book Of Love

The quest starts at the Temple of Mara in Riften. You will need to speak to a priestess, Dinya Balu, about receiving a blessing from Mara. She tells you that the blessing is only given to people that have helped Mara in the mortal realm.

You will be able to provide that help by following up on Dinya's visions, which lead to lovers in need of divine intervention. She receives her first one, sending you to Ivarstead.

Young Love In Ivarstead

Mara needs you to help Fastred, a young lady who is in love with Bassanius, a fisherman at the town. Her father, Jofthor, will not allow their romance to flourish, believing that her feelings are not real, having seen her infatuated with Klimmek not long ago. He does not want her to leave Ivarstead either. Her mother, Boti, is much more accepting and promises to shield the couple from her father.

Klimmek is the same man who will reward you for bringing supplies up to High Hrothgar, something that is best done alongside the quest the Way of the Voice.

You will have two choices, encourage Klimmek or tell Bassanius that Boti will shield them. As Mara's representative here, you decide the final outcome.

Tell Bassanius	The couple will quickly leave for Riften together.

Encourage Klimmek	Klimmek will have the courage to confess his feelings for Fastred. The two of them become a couple and stay in Ivarstead.

With either love protected, you return to the Temple of Mara and get your next request from Dinya.

Love In the Stone City

Your next person in need is none other than Calcelmo, as unpleasant as he is brilliant. He finds himself in love with Faleen, a housecarl at the Understone Keep. He is so overwhelmed by her that he cannot express his thoughts when near her.

Faleen and Yngvar may be located outside Markarth based on your decisions and progress around the Civil War and in the quest Season Unending.

He asks you to find Yngvar the Singer, a talented musician that easily sways the hearts of women. The bard tells you that Faleen has a fondness for poetry and for the low price of 200 gold, he will write a poem to make her heart melt. You must pay him.

Bring the poem too Faleen who will be so touched by the act that she writes a letter to Calcelmo and asks you to give it to him. The old researcher will be overjoyed and rush to Faleen where they confess their feelings to each other.

Once again, you will have to return to Dinya for the final request.

Love Beyond Death

Dinya will congratulate you and receive the next vision, sending you to the area around Rorikstead where you must reunite two long-dead lovers. While this task is typically impossible and morally gray, you are given an Amulet of Mara to give you the divine blessing necessary to do the Goddess' work.

The first lover, Ruki, is at Gjukar's Monument. She has been searching for Fenrig, her lover, but has not been able to find his body among the many bodies at the site. The quest marker will update, guiding you directly to him at a short distance away.

Fenrig had not perished at the battle itself, making it impossible for Ruki to find his body. After a short conversation, he will follow you. Bring him back to Gjukar's Monument.

There is no need to walk back; you can fast travel instead.

The two of them will be happily reunited and rise up into the sky. Your tasks are now complete, and you can return to the temple for a final time.

Completing The Book Of Love

With your end of the bargain held up, Dinya will make you an Agent of Mara, providing you with a blessing from the Goddess. You are also allowed to keep the Amulet of Mara for yourself.

Agent of Mara is a permanent buff that gives you 15 percent more Magic Resistance

The Amulet of Mara makes Restoration spells ten percent cheaper to cast.

EVIL IN WAITING

Starting Evil In Waiting

This quest starts as soon as you set foot inside Valthume. Valdar, the ghost of a long-dead warrior is struggling to contain the Dragon Priest Hevnoraak in his tomb. You can volunteer to defeat Hevnoraak before Valdar fades away.

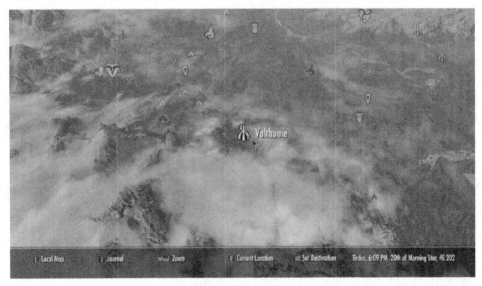

The quest will only begin when you arrive. Starting may be easier said than done as there are no quests guaranteed to bring you to its halls. If you need to find the ruins yourself, it is southeast of Markarth, deep in the cliffs.

Valdar explains that you need to gather three Opaque Vessels scattered throughout Valthume to contain Hevnoraak. He unseals the entrance, allowing you to begin.

Gathering The Vessels

The first vessel is not far from the entrance. At a large chamber, take the south exit and enter a hallway drenched with oil. Be alert for a pressure plate by the door. Standing on it drop flaming lamps on the oil.

Keep an eye out for a large grate in the middle of the large chamber. It is a trap that will force you to fight several frostbite spiders and a draugr to escape. The exit is behind a locked door with the key found on the draugr.

The vessel is guarded by a leveled draugr. Picking it up will raise a gate in the oil room, allowing you to enter Valthume Catacombs, where the second vessel is.

The catacombs have another serving of frostbite spiders and draugr, only stopping you at the gate blocking you from the vessel. There is a pull chain on the right of the gate. Defeat the draugr protecting it to open a doorway opposite the chain. You will need to walk through.

The final puzzle leading to the final vessel is a Dragon Claw wall. The Iron Claw, needed to solve it, is on a pedestal right in front of the wall. The combination, from the outer ring to the inner ring, is below.

- Dragon
- Wolf
- Bird

The last Opaque Vessel is on a pedestal in the middle of a chamber. Draugr will attack shortly after you enter. Fight them off and take the vessel to unlock the Word Wall and reach a chest containing some loot. Take the exit on the right to return to the entrance.

Defeating Hevnoraak

Valdar will instruct you to pour the contents of the vessels into a sconce and sit on a throne to awaken Hevnoraak.

The Dragon Priest will emerge in a storm of thunder and attack. While a challenging fight, it is far from the most difficult one. He will use lightning spells and summon a Storm Atronach against you.

Hevnoraak the mask, not the Dragon Priest, grants you immunity to Poison and Disease. It can be useful, especially when fighting vampires as a member of the Dawnguard, but it pales in comparison to other masks.

The quest ends when you defeat him.

SIEGE ON THE DRAGON CULT

Starting Siege On The Dragon Cult

Your journey into Forelhost starts when you arrive. You will be greeted by Captain Valmir, a High Elf soldier dressed in the same uniform as your supported faction. Yes, even if you are an elf-hating Stormcloak.

He tells you that your commander has tasked him to find a Dragon Priest mask inside the ruins. Parts of Forelhost have suffered several blockages, preventing

Valmir from getting it himself.

He unlocks the ruins and asks you to help him retrieve the mask.

Entering Forelhost

Valmir helpfully mentions that a previous attacker may have left a journal with hints on how to get past the blockages, adding the optional objective of finding Skorm Snow-Strider's journal.

The journal is already found on the route through Forelhost and is not something that you need to go out of your way to find. This objective is not difficult to complete.

When you enter Forelhost, walk straight until you reach a junction. Take a right turn, heading up and to a hallway protected by swinging axes and a dead Stormcloak. The journal is on a table at the end of this route.

You can find a linear path to the next area, Forelhost Crypt by exiting through the doorway on the left.

Getting Into The Well

The journal will mention a route into the ruins through a once-poisoned well. This well is close to the entrance to the crypt, in a small hallway, and locked behind a master-level lock.

You can, of course, pick the lock and skip the next step, if you have the skills or enough lockpicks.

Go down into the catacombs, not far from the entrance into the well. You will need to fight several rooms of dragon cultists and draugr before finding the key within a chest in a throne room.

Climb up a ramp after the throne room to return to the well. With the door unlocked, you will have to swim through some rapids and fight several frostbite spiders before you reach the final area, Forelhost Refectory.

Reaching The Mask

The Refectory is another slog through more undead defenders, interrupted by a wall blocked off with spikes. You will need to search the room for the Glass Claw, found in the middle. Picking the claw up will lower the spikes.

The Claw's corresponding puzzle wall is not far away. The combination to unlock it is below.

- Fox
- Owl
- Snake

You will need to fight a final collection of rooms before confronting Forelhost's boss, the Dragon Priest Rahgot. He, and his undead allies, must be defeated before you can loot his mask and return to Captain Valmir.

Rahgot's mask is one of Skyrim's unique Dragon Priest masks. It favors melee fighters, giving 70 points of Stamina to its wearers.

You find him in a different uniform, trying to convince an enemy soldier to enter the ruins. You will have to defeat them both to end the quest.

BLOOD ON THE ICE

Starting Blood On The Ice

The quest will only start once you pass through the graveyard after visiting the city at least four times. You will be stopped by a guard inspecting a crime scene. One of the tavern's workers, Susanna the Wicked is dead and mutilated. She is the third woman to have been murdered in the city in recent days.

Occupied with the Civil War, the guard is unable to investigate and asks for your assistance by questioning the three witnesses. You will need to talk to the three civilians surrounding the corpse. None of them provide any definitive evidence, though Helgird mentions that it is not a robbery, and Calixto Corrium claims that the killer is male.

The guard will claim that the Civil War is keeping them busy even if you have concluded the conflict.

With nothing else to go on, you can volunteer to investigate further. The guard asks you to get permission from Jorleif, the Jarl's steward, before he gives you all the details. You can find him at the Palace of the Kings. He grants his blessing quickly, hoping to bring the killings to an end.

Despite his offer to help, Jorleif rarely has good advice to give despite being indicated with a quest marker. You can ignore him for the majority of the quest.

❖ Buying Hjerim

This quest can also start if you had bought Hjerim and entered it. In this instance, you will be unable to renovate your home and will need to uncover the killer by yourself. You can jump to Searching Hjerim to complete the quest.

Investigating The Crime Scene

Speak to the guard at the crime scene with Jorleif's permission. He will be happy to give two pieces of useful information.

- The priestess of Arkay, Helgird, is conducting an autopsy on the body.

- There is a trail of blood leading away from the crime scene.

Helgird is located next to the crime scene, inside the Windhelm Hall of the Dead. She tells you that the wounds on the body could only be inflicted by embalming tools like the ones she has. Though you can accuse her of being the killer, the conversation ends there with nothing else to learn.

You will be unable to speak to Helgird if you choose to pursue the trail of blood first.

Return to the graveyard to look for a trail of blood leading from the crime scene to Hjerim, the residence of another victim, Friga Shatter-Shield. The door is shut with a Master-level lock.

Entering Hjerim

If you are unable to pick the lock, you can speak to a guard or Jorleif about getting the key. They will tell you that Tova Shatter-Shield, Friga's grieving mother, currently has it. You can ask Tova for the key to get it.

If you had killed Nilsine Shatter-Shield in the quest Mourning Never Comes, Tova will have committed suicide. You will need to break into the house of Clan Shatter-Shield to take the key from her body.

The house is unlocked during the day though remaining inside will be considered trespassing.

Tova can be found wandering between the house of Clan Shatter-Shield, the Windhelm Market, and Candlehearth Hall. She is eager to bring the Butcher to justice and hands over the key with little fuss.

Searching Hjerim

Once inside Hjerim, you need to find four items to continue your investigation

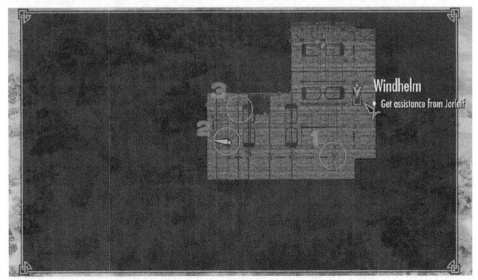

Butcher Journal #1	This journal can be found in a chest in the southeast of the house. It is on your left as you enter.
Strange Amulet	This amulet is also in the western part of the house, buried under a pile of Beware the Butcher! flyers.
Butcher Journal #2	This journal is found behind a suspicious wardrobe in the western part of the house. The wardrobe is nailed into the wall and has a false back. Interact with it to find the gristly remains of the victims and the journal.
Beware the Butcher!	These are found throughout Hjerim Hall. Make sure to take one before you leave.

With your search concluded, bring the Amulet to a guard or Jorleif. They will ask you to show it to Calixto at the House of Curiosities. He tells you that the amulet belongs to the Court Wizard and offers to buy it from you for 500 gold.

Calixto will also offer you a tour. While interesting, it is optional, and you can safely ignore it.

Bringing a flyer to a guard will have you directed to Viola Giordano. Showing her the journals and amulet will give her enough information to accuse Windhelm's Court Mage, Wuunferth. She urges you to inform Jorleif immediately.

As this is a guide, the next steps will focus on the optimal way to complete this quest.

Going to Jorleif will have Wuunferth arrested and allow a fourth murder to happen without any way of stopping it. To continue, you will have to speak to Wuunferth in the Windhelm Jail.

Stopping The Butcher

Ignore Viola's urgings and confront Wuunferth with the evidence instead. He is offended by the accusation but is otherwise cooperative. He dismisses Calixto's explanation about the Amulet and shares what he knows. He had been observing the murders and had been slowly recognizing a pattern. The Amulet is the final piece of the puzzle.

Wuunferth is found inside the Palace of the Kings Upper Floor, on the left. His room is easily identified thanks to the enchanting table.

The Amulet in question is the Necromancer Amulet. Using it as a clue, he pinpoints where the next murder will be and figures out that it will happen imminently. Go to

Windhelm's market and stand watch at night.

The Necromancer Amulet is a powerful necklace that increases your Magicka by 50 points and makes Conjuration spells 25 percent cheaper to cast. It will reduce your Health and Stamina regeneration by 75 percent.

It will not be long before Calixto appears, tailing a woman. You will have a short window to intervene before he claims his fourth victim. He will not put up much of a fight and will need to be killed.

Looting Calixto's body will get you the Necromancer Amulet if you had sold it to him and a key to his home, which will shed some light on his motivations.

Regardless of your success in saving his fourth victim, the Butcher is dead, and the women of Windhelm are safe. Return to Jorleif to end the quest. It will allow you to renovate Hjerim if you have bought it.

RISE IN THE EAST

How To Start Rise In The East In Skyrim

Head over to the East Empire Company office at the Windhelm Docks and talk to Orthus Endarios. He'll inform you that a group of pirates, the Blood Horkers, have been messing with the company in the city and have been preventing them from conducting business.

He'll ask you to find out where they're located and to take them down. He'll make sure you are well compensated for the job. This quest is made easier with some thief skills, such as sneak, lockpicking, and pickpocket, but is not necessary to complete Rise In The East.

Finding The Blood Horkers' Base

After you're done talking with Endarios, you will have to steal a logbook that can be found in the Shatter-Shields keep located to the right of the East Empire Company office. The door to the Shatter-Shields will be locked with a master-level lock preventing easy access. The door can be lockpicked but if you're having trouble, you can head to the home of clan Shatter-Shield, that's located in the northwestern part of Windhelm, and pickpocket the key off of Torbjorn Shatter-Shield.

Regardless of how you get into the keep, the logbook can be found in the back left corner of the area behind an open book that's behind a small wall. There are fewer people at the office at night so it's best to try and steal the logbook then. The logbook will reveal that the Shatter-Shields hired a pirate group to harass their competitors so they can become the main trading force in Windhelm.

Now return to the office and present the book to Endarios. He will then tell you to head to the Windpeak Inn in Dawnstar in order to find out the location of the Blood Horkers' base. Go inside the inn and talk to Stig Salt-Plank to find out the location of the base. You can either bribe him, brawl with him, or read the note on him by pickpocketing him to discover the Blood Horkers' base is located at Japhet's Folly.

Assaulting Japhet's Folly

Head back over to Endarios to report what you've learned and he will direct you to a boat that will take you to the base. Once there, talk to the East Empire agent Adelaisa Vendicci about what to do next. She'll tell you to lead their assault on the island and to eliminate the Blood Horkers' magician leader, Haldyn.

Japhet's Folly is located to the south, so jump across the ice until you can see a tower hidden by some fog in the distance which is the base. From here, you can either charge in through the docks to your left for a more direct assault, or go straight underwater to find a sea cave for a sneakier approach. Regardless of the approach, you need to head to the top of the tower where Haldyn is located and fight any enemies that get in your way.

Once you get there you will have to deal with Haldyn himself, who uses lightning bolt spells for offense, wears heavy armor for defense, and fast-healing spells to replenish his health. Be sure you have some anti-shock potions and apparel on you as well as weapons that can drain Haldyn's magicka. Thankfully though it's a one-on-one fight as he has no allies in his vicinity.

Once you slay Haldyn in Rise in the East, loot the key off of him to have an easy way out of the tower. Once outside, the East Empire will begin assaulting the island with explosive arrows, which will kill any remaining pirates and destroy nearby buildings. Keep pushing to the northeast until you get back to the boat you came on. Tell Adelaisa of the news and let her know when you're ready to return.

Completing Rise In The East And Its Rewards

You will now be returned to the Windhelm docks where Endario will thank you for finishing off the Blood Horkers as they are now no longer a threat to the company in Windhelm. He will give you gold for your troubles that will vary based on your level.

At the same time, Adelaisa will now become a follower for you should you need her She can be found at the East Empire Office in Windhelm.

THE WHITE PHIAL

How To Start The Quest

The White Phial quest starts off first in Windhelm, an alchemy shop owned by an old Altmer named Nurelion. Talk to him and he will tell you about the legend of The White Phial and his dreams of getting it.

Simply go through the dialogue. The Dragonborn will volunteer to fetch the Phial from Curalmil's final resting place at The Forsaken Cave. Nurelion will give you Nurelion's Mixture to help you get through the cave in order to reach The White Phial.

Traversing Forsaken Cave

Go to the Forsaken Cave to the west of Windhelm, as indicated on the map.

Upon entering, you'll be faced with two snow bears, before reaching the iron door.

Descend the spiral staircase past the iron door, leading deeper into the cave. Follow down the tunnel and into the main corridor. Avoid stepping on the switch on the floor, which will activate the trap ahead, and it'll shoot bolts at you.

Turning right, you'll be faced with draugrs. Keep heading straight down toward the door at the end and turn right once more. Continue down the tunnel and keep an eye out for more draugrs.

The tunnel opens up into a grand hall, but watch where you step. There are three square platforms along the floor that triggers a firetrap. Simply manoeuvre around them and get around the rubble to the entryway at the opposite end.

Avoid another switch button on the floor along the narrow corridor before reaching the door to the Forsaken Crypt.

Forsaken Crypt

Go down the short corridor and into a vast, open room. Deal with the draugr there and go into the next archway. Follow the path down the corridor, which leads you to the upper floor of the crypt. Beware the draugrs along the way.

Once you get to the upper floor of the crypt, cross over the bridge to the other side. Kill off the draugr there, and cross a wooden bridge to your right. This leads to another room.

There'll be two draugrs waiting for you. Simply lead them to the circular platform and this will activate the trap as you see in the picture above. Then go through the iron door to your left and go past the enclosed bridge.

Go down the corridor and defeat the draugrs along the way. You'll find another set of iron doors at the end. Open the doors, and you'll find a set of swinging-blades trap.

Wait for the first blade to swing by before slipping past. Do the same for the second blade. Walk further and you'll gain access to Curalmil's tomb.

Defeating Curalmil

Curalmil wields a sword and shield. He has the ability to unleash the Frost Breath Shout; as such is recommended to use weapons that deal fire damage. From using the Fire Breath Shout and Dawnbreaker to enchanting your weapons to deal fire damage, Curalmil is otherwise not too outstanding in terms of Skyrim bosses you encounter. But keep an eye out for the three average draugrs that's buried alongside him.

To be safe, bring along a follower with you (i.e. Aela the Huntress, Serana).

Don't forget to snag the Dragon Shout on the Word Wall above Curalmil's coffin.

How To Get The White Phial

After the boss fight, go down the corridor beneath the Word Wall.

At the end of the corridor is an ancient bowl. Pour Nurelion's Mixture into the bowl and a false wall will slowly descend to reveal a secret room. Enter the secret room, and you'll find the White Phial perched on a pedestal.

Skyrim The White Phial of legend

Take the White Phial and go back down the corridor. Emerging back to Curalmil's tomb, turn right, and you'll find an iron door.

Leave through the door and activate a lever that'll return you to the entrance of the cave.

Ending The Quest

❖ *Give The White Phial To Nurelion*

Return to The White Phial at Windhelm. Once you give Nurelion the phial, he will be displeased at its damaged state. He will only give you 5 septims for your trouble.

❖ *Speak To Quintus Navale*

Although this is optional, speak to Quintus Navale in order to obtain a greater reward. Quintus will say you deserve more than what Nurelion's given you. He will then hand you 500 septims, thus ending the quest.

CIVIL WAR

JOINING THE LEGION

How To Join The Imperial Legion

There are two ways to get the prompt to join the Legion.

1. Side with Hadvar in Unbound. He will suggest joining up with the Legion after you escape Helgen.

2. Speaking to Imperial Soldiers patrolling Skyrim or at their camps.

This will add a Miscellaneous quest to Join the Imperial Legion with a quest marker leading to Castle Dour in Solitude.

You can still join the Legion even if you sided with Ralof in your escape.

At Castle Dour, you speak to General Tullius about joining the Legion. Impressed by your survival at Helgen, he asks you to report to Legate Rikke, the female commander by the war table. She'll arrange a test; remove the bandits living in Fort Hraggstad in the mountains west of Solitude to earn her approval.

Clear out Fort Hraggstad

The ability to clear out a fortress solo is, undoubtedly, a valuable skill for any aspiring military recruit. Take the road west of Solitude and at the statue of Meridia, take a right, going north. At the split in the road, continue going west until you find the Fort.

The Fort itself is home to a large group of bandits. How you clear it is up to you, though with a lookout at the front gate, any direct approach will likely be noticed. While convenient, choosing to simply assault the Fort will be dangerous thanks to its design. The bandits will be able to attack you from all directions as soon as you reach the central courtyard. A stealthy approach would be much safer.

You can infiltrate Fort Hraggstad by entering through a hole in its northern wall by a collapsed tower.

In addition to the bandits patrolling the outer walls, you will need to enter the inside of the Fort to flush all of them out. You need to use the two entrances in the central courtyard to reach the two interior areas to wipe out the remaining bandits.

These interior areas have stronger bandits inside, making them a tougher fight.

With all the bandits slain, return to Legate Rikke at Castle Dour to formally join the Legion and receive your next instructions, participating in the mission to seize the Jagged Crown.

You will also have the opportunity to speak to Beirand, Solitude's Blacksmith to receive a full set of armor. The specific set depends on your preference.

Light	Imperial Light Helmet, Imperial Light Armor, Imperial Light Bracers, Imperial Light Boots, and Imperial Light Shield
Medium	Imperial Light Helmet, Studded Imperial Armor, Imperial Light Bracers, Imperial Light Boots, and Imperial Shield
Heavy	Imperial Helmet, Imperial Armor, Imperial Bracers, Imperial Boots, Imperial Shield

JOINING THE STORMCLOAK REBELLION

How To Join The Stormcloak Rebellion

Broadly speaking, there are two ways to get a Miscellaneous quest "Join the Stormcloak Rebellion"

1.   Side with Ralof on your escape from Helgen in Unbound.

2.   Speak to Stormcloak soldiers on patrol or at their camps

In both instances, you will be prompted to go to the city of Windhelm and to speak with Ulfric Stormcloak to join the Rebellion. He will be in The Palace of the Kings, speaking to his right-hand man Galmar Stone-Fist.

Remembering Helgen, Ulfric acknowledges that you might be an asset to his army and asks you to speak with Galmar about joining. The old warrior sets a test for you kill an ice wraith to prove that you have the strength to join the Stormcloaks. This starts the full quest, Joining the Stormcloaks.

This should go without saying, but joining the Stormcloaks will stop you from joining the Legion.

Where To Kill The Ice Wraith

Your initiation will take you to the Sea of Ghosts, in the frozen waters north of Skyrim. There, you need to find your way to Serpentstone Isle to fight the ice wraith. The Isle itself is in the middle of the sea and cannot be accessed without swimming across.

The closest city is Winterhold. There will be a safe path down the cliff on its east side that should provide you with a route to Serpentstone Isle though there is no road there.

If you are struggling to kill it, Galmar will provide you with an Elixir of Resist Cold and three Ice Wraith Banes, a unique poison that will deal extra damage to ice wraiths. These items will make it much easier to defeat the ice wraith.

There will be a slope on the north side of Serpentstone Isle that you can use to reach the ice wraith. It will turn aggressive as soon as it notices you, immediately starting combat. The fight itself will be a challenging fight against a single enemy. As the name implies, the ice wraith is weak to fire and deals frost damage that saps your stamina.

Serpentstone Isle takes its name from the Serpent Standing Stone, a magic stone that grants you the power to paralyze enemies once a day.

Once you defeat it, return to Windhelm and speak to Galmar. He will be impressed with your performance and allow you to join the Stormcloak Rebellion, giving you a uniform to look the part. Your first mission with them is to retrieve the Jagged Crown to give Ulfric legitimacy as the true High King of Skyrim.

THE JAGGED CROWN

Starting The Jagged Crown

This quest starts immediately after you officially join the Imperials or the Stormcloaks. Either Legate Rikke or Galmar Stone-Fist will give you your orders to assist with this mission.

Korvanjund is in a relatively central part of Skyrim. A fast way there would be to travel north from Whiterun, following the road until you find your officer and their unit waiting to start the assault. They will tell you that the enemy has already reached the ruin and that you'll need to fight through them. Tell your officer when you are ready to begin.

You will be greeted by whoever you escaped Helgen with, either Hadvar or Ralof provided that you joined their side.

Clearing Korvanjund

You will be led into the ruins where a large group of enemy fighters will be waiting for you. With the long, narrow spaces and verticality offering vantage points, enemy archers can be especially deadly. Make sure to advance with your unit to avoid being picked off alone.

You don't actually have to listen to your orders. As long as you kill the enemy, you'll do fine.

With the outside cleared out, the next step is to enter the ruins. You will be ordered to start stealthily to get into position. With numbers on your side, the enemy fighters will be quickly overwhelmed. After leaving two fighters to guard the entrance, the rest of your unit goes in deeper.

At the second room, your leader will stop the advance, fearing that there is an ambush on the other side. They order you to find a way through and ambush the enemy while promising to back you up if fighting starts. Go up to the second floor and get on the bridge. You can fire arrows at the enemy, catching them by surprise. The rest of your unit will rush in at the first sign of fighting.

Alternatively, you can just walk right into the ambush. This will be more difficult, but it is entirely survivable.

After that, you will reach a room with a single enemy soldier and a fallen draugr. The undead Nords will be returning as an enemy once again, to the discomfort of your unit. Next, your officer will lead the unit into Korvanjund Halls.

❖ How To Solve The Korvanjund Halls Dragon Claw Puzzle

Two enemy soldiers will be waiting at the entrance, fighting you immediately. They will be guarding a dragon claw puzzle like the one at Bleak Falls Barrow. To get through, pick up the Ebony Claw, on the ground just in front of the puzzle. The combination, from top to bottom, is Wolf, Butterfly, Dragon. Put the Claw into the relevant socket to open the door.

There is a treasure room guarded by swinging axes before you reach the puzzle. You can use Whirlwind Sprint to get through without getting sliced to ribbons.

❖ Finding The Korvanjund Halls Lever

In the next room, your officer will be unsure of where to go next and will order you to find a way through. You need to go through the doorway on the far right of the entrance, head up to the second floor, go across the bridge and find the lever by an urn. Pulling it will open the way forward and awaken the sleeping draugr. They will come from all sides in an ambush. Though they will be quickly defeated.

From here, you will be able to reach the final room in Korvanjund Crypt. There are just three enemies here, in the form of three powerful draugr. The one in the center is wearing the Jagged Crown and is the target of the operation. Be mindful of its shout attacks and powerful weapons as you beat it down and take the Crown for yourself.

You'll be ordered to return to your leader at your faction's capital to end the quest. This will also begin the next quest, Message To Whiterun, which precedes the first major battle of the war, The Battle For Whiterun. This time though, the experience will be very different depending on which side you're on.

If you are unsatisfied with your leadership, you can deliver the Crown to the enemy instead. This will allow you to switch sides without needing to complete the recruitment quest.

THE BATTLE FOR WHITERUN

Starting The Battle For Whiterun

The quest starts immediately after completing Message To Whiterun. You will be ordered to go to Whiterun to deliver an item to Jarl Balgruuf. Jarl Balgruuf will speak with his advisers before deciding to side with the Empire. Complete the remaining quest objectives to begin the battle.

Imperials	Deliver a message from General Tullius. Jarl Balgruuf then asks you to give Ulfric Stormcloak an axe.
Stormcloaks	Give Jarl Balgruuf an axe and return to Ulfric Stormcloak

Fighting The Battle For Whiterun With The Imperial Legion

Siding with the Empire will have you manning the walls against the Stormcloaks. You will have two objectives.

1. Defend Whiterun

2. Kill all the Stormcloaks

You start at the gatehouse facing the Whiterun Stables. After a short speech from Legate Rikke, the assault begins. Stormcloaks will rush the barricades and attempt to destroy them. You can use ranged attacks against the attackers or jump in front of the barricades to defend them.

You will not fail the quest if either objective is lost. Don't worry if they destroy the barricades or lower the drawbridge

The Stormcloaks will attempt to lower the drawbridge next, running through the wooden structure directly across from the entrance to do so. Though with the fortifications on their side, the defenders will win.

The quest will automatically update as you thin out the Stormcloaks, letting you know how much longer you need to hold out for.

At the end of the battle, you can listen to Jarl Balgruuf give a speech to the defenders where he thanks them for their bravery. He will also grant you permission to buy a home in Whiterun if you have yet to earn the right.

To finish the battle, return to General Tullius who will promote you and give you a leveled weapon. With your new rank, you begin the quest Reunification of Skyrim and receive your orders, giving you the next province and camp to report to.

Fighting The Battle For Whiterun With The Stormcloak Rebellion

As the attackers, the Stormcloaks need to break through the barricades protecting the city to force Jarl Balgruuf to surrender. Starting at a Stormcloak Camp by city, you will be following the Stormcloaks through the city and all the way to Dragonsreach.

There will be wooden barricades preventing you from progressing. Attacking will destroy them and allow you to proceed.

The first barricades will be at the city gates, near the stables. Try to break through quickly; enemy combatants will shoot arrows at you as you climb up. You will need to lower the drawbridge once you're in. To do so, go across and onto the wooden platforms, following them into the gatehouse. The defenders will flee after you lower it.

Chase them into the city. From here, how you get to Dragonsreach is up to you. However, you may wish to follow the Stormcloaks through the market, taking the nearby stairs up to Gildergreen. Going straight ahead from here will allow you to reach Dragonsreach quickly.

You can break through barricades to make your own way through the Plains District on your way to Dragonsreach.

All that is left now is to defeat Jarl Balgruuf in combat. With his steel armor and Irileth backing him up, the Jarl will be a difficult opponent to fight. It's best to let him fight Galmar and Ralof as you pick off any allies he might have. Once you batter him down, he will surrender, ending the battle.

Vignar Gray-Mane arrives, having an argument with Balgruuf before he takes control of the city. You will be asked to return to Ulfric in Windhelm. He will promote you to Ice-Veins and give you a sword based on your level. The quest Liberation of Skyrim will begin, giving you the next province and camp to report to.

There is a common bug where you cannot report the victory to Ulfric and Whiterun will remain under attack even after you win. This is usually caused by fast traveling to Windhelm directly.

If you encounter this, reload a save prior to defeating Balgruuf and walk to Windhelm. Whiterun should return to the map as a fast-travel location, signaling that you are in the clear.

RESCUE FROM FORT NEUGRAD

Starting Rescue From Fort Neugrad

You get this quest when you report to Galmar Stone-Fist at the Falkreath Stormcloak Camp. Ulfric will send you there during Liberation of Skyrim, the overarching quest that tracks your progress through the Skyrim Civil War.

You will not do this quest if Falkreath was traded to the Stormcloaks in Season Unending.

You will need to link up with Ralof and his scouts near Helgen, where the game first started. There, he will suggest that you infiltrate Fort Neugrad, taking a hidden path to enter the prison directly and free the prisoners without being detected. He will come to help you when you reach the courtyard.

If you don't fancy sneaking through the Fort, you can simply charge right in though you will still have to free the prisoners after you kill all the Imperials.

Sneaking Into Fort Neugrad

The path Ralof suggests involves sneaking down a hill south-east of his position, just under the walls of the Fort. Make sure to extinguish any light sources to make the approach easier.

You will have to dive into the lake, looking for a hidden cave on the north side, at the foot of the Fort. Swim right inside to enter Fort Neugrad, right next to the cells.

Once inside, keep your eye out for a patrolling guard by the prison cells. Kill him quietly to stop him from calling for help. Taking his key will allow you to free the imprisoned Stormcloaks who will promptly arm themselves and help you fight the Imperials.

Being discovered here is not the end of the world. You'll still be able to capture Fort Neugrad, though it might involve you fighting alone for a time.

There will be another guard patrolling the area above, though he will not pose a significant threat. When all of the prisoners are free and armed, follow the quest marker to find the door leading to the courtyard. Ralof will come to help as soon as you exit.

Capturing Fort Neugrad

With Ralof's help, the Fort is easily captured, though breaking the wooden barricades will help him help you more quickly. Keep an eye out for archers as you fight Imperial soldiers. You may have to sweep interior areas to capture it.

Once the Fort has been cleared out of the Empire's lapdogs, you can return to Ulfric to announce your victory and get a promotion.

You can find the Fort Neugrad Treasure Map on the roof. It leads to a treasure chest.

RESCUE FROM FORT KASTAV

Starting Rescue From Fort Kastav

You will get this quest as you progress through Reunification of Skyrim, the Imperial path for the Skyrim Civil War after some early victories elsewhere.

You might not get this quest if gave Winterhold to the Empire in Season Unending.

General Tullius will order to you join the effort to retake Winterhold by sending you to the Winterhold Imperial Camp, where you will start the quest after speaking to Legate Rikke.

Sneaking Into Fort Kastav

You will meet Hadvar when you link up with the scouts. He will give you a lay of the land and provide you with an infiltration route. He will also reassure you that he will help you if you are detected.

It's best to approach the scouts from the south-west of the fort. That way, you will not pass close to the Fort on your approach.

To sneak into the Fort, walk straight at a grate, directly in front of the scouts. Though it takes you up a slope in full view of the Fort, there are few guards, and should not be too challenging to execute.

Once inside, you will have to deal with three Stormcloak guards. It is best to kill them first as the prisoners will be unarmed and helpless immediately after you free them. The guards have the following patrol routes.

By the infiltration point	There will be a single guard that patrols the rooms just past the infiltration point. Failing to silently kill him will attract other guards.
Patrolling the prison	This guard walks a path through the prison cells and near the foot of the stairs.
Sitting above the cells	A third and final guard is waiting above the prison cells by a table. You can snipe him as soon as you enter.

All the guards have a Fort Kastav Prison Key, which you use to open the cells, though you could pick the lock. The prisoners will rush to a chest to arm themselves and will wait for you to enter the courtyard and remove the Stormcloaks.

Capturing Fort Kastav

As soon as you set foot in the courtyard, the battle for Fort Kastav will begin. Like the other Forts, you will have to reduce the Stormcloak forces to 0 percent to win. There is nothing particularly remarkable at the Fort that distinguishes it from the others.

If you're not the sneaking type, you can rush in and fight the Stormcloaks without sneaking. You will still complete the quest after freeing the prisoners.

Talking to Hadvar after capturing the Fort will have him send you to General Tullius to complete this stage of the Reunification of Skyrim.

COMPELLING TRIBUTE

Starting Compelling Tribute

This quest starts when you are working to capture The Rift or The Reach, depending on which side you are on. Your commander will explain that you need to break into the Hold's Keep to blackmail the local Steward. The blackmail material will be hidden in the Steward's quarters, forcing you to slip past patrolling guards.

You will be able to freely enter the restricted areas if you are Thane of that Hold.

Once blackmailed, the Steward will tell you about a wagon full of loot that you can intercept and capture. In doing so, you gain the opportunity to overpower a nearby Fort, capturing the Hold for your side.

Intercepting The Wagon

Your commander had convienetly sent a team along the wagon's route and asks you to link up with them. Follow the quest marker to find them by the road. You will meet either Hadvar or Ralof who will suggest a plan that has you sneak up to the wagon, stealthily eliminating enemy soldiers as you approach.

The backup provided by your allies will make the quest much easier. They will act as soon as combat starts, making them reliable.

Following his plan is safer thanks to the men backing you up. However, you can opt to ignore the plan and go in without sneaking. This will alert all the guards but may suit you better.

After dispatching all the guards, you will be able to pick the lock on the carriage, allowing you to loot the cargo, if you need the gold.

Compelling Tribute For Imperials

This quest starts when you need to bring Riften under Imperial Control. After reporting to Legate Rikke at the Rift Imperial Camp, you are sent to Mistveil Keep in Riften to find blackmail material for Anuriel.

❖ Finding The Evidence At Mistveil Keep

The evidence, Incriminating Letter, is found in Anuriel's room, inside the Keep behind the Jarl's throne. It is through a fairly narrow hallway that lacks any hiding spaces. However, you will be able to slip inside if you time your entry well to avoid the Guard patrols.

Mistiming your infiltration will not have any significant consequences, so long as you quickly leave.

Confronting Anuriel with the Letter will have her reveal that there is a wagon of gold coins heading for Windhelm. An easy persuasion check will net you a personal bribe as well.

Sharing this information with Legate Rikke will have her send you to join a scouting team that's already patrolling the area. You will meet Hadvar, now a commander in the Legion, with the scouts.

❖ The Battle For Fort Greenwall

The final step before capturing the Hold is to take Fort Greenwall from the Stormcloaks. The Imperial forces will be waiting for you at the south-east of the fort. You can attack from the other side to start a two-pronged assault. Fight the

Stormcloaks until you deplete their numbers and capture the Fort.

Compelling Tribute For Stormcloaks

You will get this quest when reporting to Galmar Stone-Fist when liberating the Reach from Imperial rule. Galmar explains that you need to sway Raerek, Steward of the region, to the Stormcloaks' side. With the rumor that the man worships Talos, an act outlawed in Skyrim, you need to find dirt on him and blackmail him.

❖ Finding The Evidence At the Understone Keep

You will have a quest marker pointing right at the cupboard hiding the evidence. It is locked behind an adept-level door and is almost always guarded. You will need a high sneak skill and good lockpicking skills to break into the room.

If you are struggling to get into the room, Raerek will open the door when he sleeps there for the night, giving you an opportunity to break in. Take Raerek's Inscribed Amulet of Talos from the cupboard. With the Amulet in your inventory, you will be able to blackmail him.

If you are really struggling, you can rush into the room and steal the amulet. If no one sees you take the amulet, they will merely demand that you leave. You will be able to confront Raerek later.

Raerek will tell you about a wagon filled with Silver and weapons heading to Solitude. You can also attempt to persuade him to give you a personal bribe as well. With this information in hand, you can return to Galmar. He will send you to meet a team of scouts before you capture the wagon

❖ The Battle For Fort Sungard

With the Imperial presence weakened, Galmar orders you to take on Fort Sungard to claim the region for the Stormcloak rebellion. There are no special requirements to claim the fort.

A FALSE FRONT

Starting A False Front

This quest begins when you speak to your commander when you begin the campaign to take on the next Hold. You will need to locate a courier traveling between two Inns and steal their orders to return to your commander. You will then have to bring the modified orders to the enemy commander, laying the groundwork for a subsequent assault.

To find the courier, you will need to visit either of the Inns to find hints about their location. Speaking to the Innkeepers will yield quick results, though you will have a few options for going about it.

Persuade	You will need a high Speech skill for this to succeed
Bribe	It will cost you about 500 Gold to pay the Innkeeper for information
Intimidate	This is likely to succeed.

Making the Innkeeper spill the beans will provide you with a quest marker for the courier themselves. The most convenient course of action would be to kill the courier out in the wilds where there are no witnesses to the murder.

If you are skilled at pickpocketing, you could wait at the inn for the courier and steal the Orders,

Each Faction's version diverges around this point. Refer to your faction's version of the quest to continue.

A False Front For Imperials

Unless Dawnstar is under Imperial Control after Season Unending, this is the first objective in the quest, Reunification of Skyrim. The quest itself will begin after reporting to Legate Rikke in the Pale Imperial Camp.

The Legate explains that she wants you to intercept a Stormcloak Courier that ferries messages from Dawnstar to Windhelm. She's identified that the courier stops at the Nightgate Inn, located in the snowy area near Dawnstar, and Candlehearth Hall, in Windhelm.

Nightgate Inn is in an inaccessible location and is far from other fast travel locations. Due to being in a major city, Candlehearth Hall an easier Inn to search.

That said, the lack of residents makes it easier to kill the courier without anyone realizing.

On your delivery, Legate Rikke will modify the documents to contain incorrect information. You will have to travel to The White Hall, where the Jarl of Dawnstar resides. Look for the local Stormcloak Commander to deliver the orders. If you are wearing Stormcloak armor, you will get some unique dialogue.

❖ The Battle For Fort Dunstad

The final stage of the quest is to capture the weakened Fort. This will be a direct assault through some wooden barricades. There is no need to enter any interior locations. Staying outside and slaughtering enemy combatants will allow you to win. Once the enemy percentage reaches 0%, the battle is won.

A False Front For Stormcloaks

This quest starts for Stormcloak players when it's time to take Hjaalmarch and its capital, Morthal, from the Imperials. This is skipped if the Hold was given to the Stormcloaks in Season Unending. Galmar Stone-Fist will give you the quest when you report to the Hjaalmarch Stormcloak Camp.

You will need to intercept a courier that Galmar knows is traveling from the Four Shields Tavern in Dragon Bridge to Frostfruit Inn at Rorikstead. Both areas are smaller settlements, and it would be easy to follow the courier out into an empty space to kill them without any witnesses.

Being closer to the camp, Dragon Bridge is an easier location to conduct the operation.

With the orders in hand, return to Galmar to have them modified. You will need to deliver the modified orders to the Imperial Commander at Highmoon Hall in Morthal will allow you to fool the Imperials into weakening Fort Snowhawk and make the next stage of the liberation of Hjaalmarch possible.

❖ The Battle For Fort Snowhawk

Galmar will order you to meet the men preparing to attack Fort Snowhawk, west of Morthal. You will find them waiting for you by a cliff near the fort.

Located near a river and with archers situated on its many battlements, Fort Snowhawk's defenders are able to fire arrows at attackers with impunity. It is especially important to break the barricades to let the main Stormcloak force before the Imperials pick them off. You will not have to fight in any interior areas for this battle.

BATTLE FOR WINDHELM

The Battle For Fort Amol

Before you can start the fight against Windhelm, Legate Rikke will send you to capture Fort Amol, a fortification just north of Ivarstead. This battle will be similar to the other ones that you have fought before. You need only deplete the Stormcloak troops to win the day.

Breaking the wooden barricades at the entrance to the Fort will help the Imperials break in more quickly while minimizing the risk from Stormcloak archers.

Once you have captured Fort Amol, return to the Eastmarch Imperial Camp.

The Battle For Windhelm

You will be immediately sent to Windhelm on your return. Go to the Windhelm Stables and run across the bridge to listen to General Tullius give his speech. As soon as the speech ends, enter the city.

Windhelm has been devastated by bombardments, with rubble blocking its streets. The paths that do remain are barred with wooden barricades. You will need to fight through groups of Stormcloak soldiers to reach the Palace of Kings and fight Ulfric.

Head west, breaking through the barricades leading to the forge, and turn north to reach the graveyard. Continuing north through the residential area will get you to the Palace of Kings. A final group of Stormcloaks will block your way. Enter the Palace as soon as you can.

Defeating Ulfric and Galmar

General Tullius and Legate Rikke will join you as you stare down the two leaders of the Rebellion. After making initial demands and offers of surrender, Ulfric will choose to fight to the death. This starts a final battle in the throne room. You need to kill Galmar Stone-Fist and defeat Ulfric to bring the battle to its conclusion.

Both Stormcloak leaders are powerful warriors in their own right and will be challenging to defeat. This is complicated by the fact that Ulfric can use the Voice against you.

Your allies will likely attack Galmar first, leaving Ulfric for you. Denying Ulfric the opportunity to use the Voice against you is essential. Use power attacks, shield bashes and other stunning attacks to keep his Shouts at bay. Your allies cannot die and can be used as a form of protection against Galmar.

After being defeated, Ulfric will ask to be killed by the Dragonborn. You can grant his wish or let Tullius end him. With the leader of the rebellion slain, Tullius and Rikke will leave the Palace to give a speech.

In completing this quest, you have ended the Civil War with the Imperials winning.

BATTLE FOR SOLITUDE

The Battle For Fort Hraggstad

After being dispatched by Ulfric to the Haafingar Stormcloak Camp, you are sent to help capture Fort Hraggstad, in the mountains west of Solitude. Fast travelling to the Thalmor Embassy will provide you with a fast route to the Fort.

By this point, you should be familiar with capturing Forts from Imperial lapdogs. You will not have to enter any interior areas to capture the Fort. Simply deplete the Imperials to capture the area.

The Battle For Solitude

As soon as you report your success to Galmar, you are sent to the gates of Solitude to join up with Ulfric and the assaulting Stormcloaks. After a speech, it's time to enter the city.

Your objective is to reach Castle Dour, where General Tullius is waiting. As you make your way through Solitude, you will have to fight against Imperial Soldiers waiting throughout the city. You will not be fighting alone as you follow Ulfric and Galmar alongside the Stormcloaks throughout the attack.

Ulfric will be using The Voice to disarm Imperial Soldiers throughout the fight. Do not be alarmed by another warrior using Dragon Shouts.

With a prior bombardment damaging the city, you will not be able to take a direct route to Castle Dour. Head east, through the Solitude Market until you reach the street that leads to the Blue Palace to the south and Castle Dour to the north. Turn north to close in on the General, breaking through the last of his men.

Defeating General Tullius

Entering Castle Dour will put you in the final stage for the Battle of Solitude. With Galmar barring the door, the three of you confront the two Imperial commanders, General Tullius and Legate Rikke. Though Ulfric and Galmar make it clear that they are only after the General, the Legate chooses to fight regardless.

While stronger than your average legionary, neither commander is particularly remarkable and can be easily defeated. Thanks to your two companions, it will be easy to split the two commanders and deal with them one at a time. The battle ends when Tullius is forced to his knees and Rikke is slain.

You will get the choice to end Tullius yourself or to let Ulfric do it. There are no consequences for either option save for narrative reasons.

DAEDRIC ARTIFACTS

DISCERNING THE TRANSMUNDANE

Starting Discerning The Transmundane

This quest starts when you search for an Elder Scroll locked deep within Dwarven ruins. You may be prompted to start searching after encountering Paarthurnax at the Throat of the World or during the Dawnguard Quest Scroll Scouting. Both quests point you to the College of Winterhold where you speak to Urag gro-Shub about finding the Scroll.

He offers the limited extent of the College's resources on the topic, including the book Ruminations on the Elder Scrolls. The tome is nearly indecipherable and is confusing enough to start Discerning the Transmundane.

Questioning Urag will start your journey to the frigid ice floes far to the north of Winterhold where you find Septimus Signus' Outpost. The half-mad scholar will help you find the Elder Scroll and, more importantly, give you a Blank Lexicon. You will need to bring the Lexicon to the Tower of Mzark to continue.

The actual journey to reach The Tower of Mzark is both long and difficult. We covered exactly how to get there in our guide for the main story quest Elder Knowledge. This guide also covers how to complete the puzzle that lets you get the Runed Lexicon.

Now, you'll need to head back to Signus. Luckily, there's a lift in the tower you can use instead of heading through Blackreach the long way. It was blocked from the inside, so pull the lever to unlock it.

Harvesting Elf Blood

Bring the Runed Lexicon back to Septimus to continue the quest. He will tell you that the locked box needs Dwarven blood to be opened. Without any living specimens, he asks you to bring him a cocktail of elven blood to allow him to fool the lock.

Once you've returned to the Outpost, the game will require you to be level 15. If you're not, Signus will dismiss you, and, when you reach level 15, you'll receive a letter from a messenger.

On your way out to begin this process, you'll be stopped by the Wretched Abyss. This is the Daedric Prince, Hermaeus Mora. He'll tell you that Signus will soon outlive his purpose, and offers you the chance to replace him. It doesn't matter what you decide here — it won't change anything in the quest.

It's time to harvest the blood that Signus needs. You'll need some from each of the five elven races: High Elf, Wood Elf, Dark Elf, Falmer, and Orc. Some locations, like Alftand, Liar's Retreat, Nightcaller Temple, Rannveig's Fast, and Silverdrift Lair contain several kinds of elves, each missing one or two varieties.

High Elf	Altmer can be found in Northwatch Keep, Halted Stream Camp, Shrine of Talos Massacre, Thalmor Embassy, and Uttering Hills Cave.
Wood Elf	Bosmer can be found in Bthadamz and Autumnshade Clearing.
Dark Elf	Dunmer are found at Drelas' Cottage, Evergreen Grove, Knifepoint Mine during "Beothiah's Calling," Halted Stream Camp, Reeking Cave, and Stonehills.
Falmer	Falmer are found in almost any Dwemer ruins or the Frostflow Lighthouse.
Orc	Orsimer can be found in Cracked Tusk Keep or Halted Stream Camp.

You will need to interact with each body to extract its blood. It opens a menu that gives you the option to choose between looting the body or extracting the blood.

There is no quest marker to help you find each sample. There is no time limit to collect the samples, meaning that you can journey across Skyrim and slowly collect the samples as you encounter them.

Once you have gotten all the samples, return to Signus at the Outpost. He'll unlock the box and reveal the treasure within.

What To Pick For The Oghma Infinium

Inside this room, the mage will find that he's been wrong the whole time. The Heart of Lorkhan is not contained within, but rather, a strange book sits on the pedestal.

However, he doesn't have much time to sort it out, as Hermaeus Mora decides he no longer needs him. He then disintegrates Signus. Head up to the pedestal with the book to find it's called the Oghma Infinium.

You'll be rewarded for finding the book, and given a choice between four options, three of which will level up a series of skills: The Path of Might, The Path of Shadow, and The Path of Magic, as well as the option to not read it at all. Choosing the latter will allow you to make the decision later in the game.

The Path of Might	Your Warrior skills will increase by five points.
	Smithing, Heavy Armor, Block, Two-Handed, One-Handed, and Archery.
The Path of Shadow	Your Thief skills will increase by five points.
	Light Armor, Sneak, Lockpicking, Pickpocket, Speech, and Alchemy.
The Path of Magic	Your Mage skills will increase by five points.
	Conjuration, Restoration, Alteration, Enchanting, Illusion, and Destruction.

When you try to leave, you'll be confronted by the Wretched Abyss again. You can agree to be his champion or refuse. Neither of these choices has any real effect, so it's up to you.

Your interaction with Hermaus Mora will be influenced by your interactions with him during the Dragonborn expansion, if you have started it.

After this conversation, the quest ends, and the Oghma Infinitum will disappear from your inventory.

THE CURSED TRIBE

Starting The Cursed Tribe

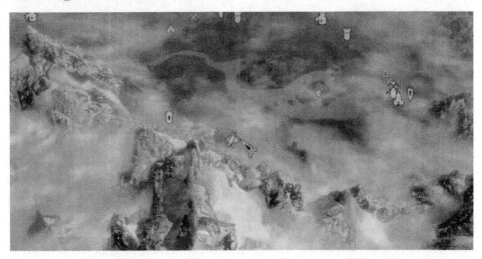

The stronghold in question is located in The Rift. The stronghold isn't too far off the cobblestone path, and you should be able to spot the marker for it on your compass. It's nestled into the foot of the mountains. As you approach, you'll notice the local orcs are in the process of fighting off a giant. Help them finish the battle, and you'll be prompted into a dialogue by one of the residents.

To get there relatively easily, you can simply travel to Riften first and then exit the city through the back gate which is located between Honorhall Orphanage and Mistveil Keep. Follow the cobblestone road that passes next to Lake Honrich, until you reach the lake's southwest area.

Atub will tell you that the orcs of Largashbur are in a bit of a pinch. Malacath has cursed their tribe's leader, Chief Yamarz, which in turn has inspired the local giants to attack their stronghold and take over Malacath's shrine. To mend the situation, Atub will request you find two ingredients to offer to Malacath in order to commune with him.

Gather The Ingredients

The two ingredients Atub asks for are difficult to gather at the best of times. They are sourced from especially dangerous enemies that are not easily found.

If you are a member of the Companions, you can find both ingredients in the Harbinger's room, on the shelves surrounding the table with maps.

❖ Troll Fat

The first ingredient Atub will request is Troll Fat. If you don't have any at hand or haven't run into a troll, you can just buy some from an alchemy merchant. Falkreath also has two locations with Troll Fat conveniently out in the open: Grave Concoctions and Corpselight Farm.

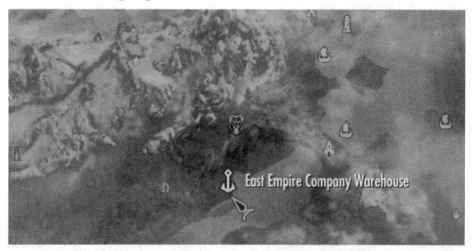

As a rare alchemical ingredient, you might want to have several portions of Troll Fat. You can find it at the East Empire Company Warehouse next to Solitude, where you can find over a dozen portions throughout. This is the most certain way of getting one if you'd rather not pay up at an alchemist's store.

❖ *Daedra Heart*

The second ingredient, a Daedra Heart is much trickier, especially if you're attempting this quest early into the game and haven't had the chance of slaying any Daedra. Whiterun's Arcadia's Cauldron will sometimes sell it, but it's generally uncommon.

The quest The Black Star will allow you to harvest Hearts from the Dremora found at the end of the quest while getting you another Daedric artifact.

Alternatively, any mages who have entered the College of Winterhold can get on Daedra Heart from Enthir once a day. There are other quests located in Dawnstar, like Pieces of the Past related to Mehrunes Dagon, and Waking Nightmare related to Vaermina. These quests will be difficult at lower levels. However, if they are available, both of them will have Daedra Heart as loot.

Return For The Ritual

Once you've gathered your two ingredients, return to Largashbur and speak with Atub once more. She'll take the ingredients and begin communicating with Malacath, with Chief Yamarz.

It turns out that Malacath isn't pleased with the state of leadership in the stronghold and demands that Yamarz clears his shrine in Fallowstone Cave and bring the giant leader's club as an apology. Unfortunately, your participation does not end here.

Yamarz believes you're to blame for all this and must accompany him to Fallowstone Cave to see this whole deal done.

Head To Fallowstone Cave

Fallowstone Cave isn't too far from Largashbur. You can choose to either travel with Yamarz or meet him there. Again, the simplest way to get there is to head to Riften first. This time, start from the front gate where the stables are located and head towards the northeast, into the mountains.

Fallowstone Cave will have a variety of enemies, from bears to trolls. However, giants are the main challenge here and considering they'll always win a melee fight against the Dragonborn, it's better to keep your distance and use a powerful bow or magic instead.

If you have a follower with you, it's also a good idea to ensure they're tanky enough to take most of the damage for you with Yamarz while you focus fire the giant.

Once you clear out the cave, you can enter Giant's Grove, which is a special location at the end of the area, where the final boss fight of the quest will take place.

Dealing With The Giant

It's no secret by now that Yamarz is a bit of a coward. So much so that as you enter Giant's Grove where the leader of the giants is located, he'll offer gold in return for slaying the giant for him.

If you agree, Yamarz will turn on you after you've defeated the giant, and you'll be forced to kill him as well. If you refuse, however, Yamarz will approach the giant and die almost instantly.

It is most convenient to refuse the gold to avoid fighting him later   .

Either way, it's up to you to slay the leader of the giants. As with regular giants, the best strategy for any build is to have a tanky follower take some of the damage for you while you attack it from afar. Eventually, however, your follower will be downed, and the giant will turn its focus on you.

If neither of those is an option, there's a small spot located on the right side of Malacath's shrine where you can stand safely without being hit. It's possible to damage the giant from this location. It can be a bit tough to access since it involves lodging yourself between a rock and the shrine.

Mages will have a small advantage against him, provided they have the Impact perk, which allows Destruction spells like Ice Spike to stagger and slow down the giant enough to keep it at bay. Powerful archers will also find this fight straightforward, so long as they keep their distance.

Getting Volendrung

With the giant defeated, you need to bring Shagrol's Hammer back to Largashbur. Upon your return, Malacath's voice will be heard again. With Yamarz's death, the tribe is offered another chance of proving itself, with a brand-new leader. Malacath will also ask you to place the hammer on the shrine.

The hammer transforms into Volendrung. It is equivalent to a Dwarven

Warhammer and deals 25 points of damage. Its enchantment lets it drain a staggering 50 points of Stamina it leeches off of any target it comes in contact with. This makes it particularly powerful in the hands of tanky Dragonborn builds.

THE BLACK STAR

Starting The Black Star

To start the quest, visit the Shrine of Azura. You can visit it of your own volition or get a miscellaneous quest from an innkeeper by asking for rumors. You can find it by heading to Winterhold and going south. The Shrine of Azura is located on top of a mountain and can easily be seen from the ground by most players.

There is a dragon location nearby, so follow the path and steer clear of any extra battles. Once at the shrine, climb up the steps to speak with Aranea Ienith, who will give you a vague description of an elven mage.

Though the quest marker brings you right to Nelacar, you can ask staff members from the College or the local innkeeper about him. They will give hints about his backstory if asked.

Head back to Winterhold to find Nelacar at The Frozen Hearth. Nelacar will be reluctant to tell you about the Azura's Star, so you will need to persuade him using speech, gold, or intimidation. After he finally speaks to you, he will tell you to visit Ilinalta's Deep near Riverwood.

Finding Azura's Star

When you reach Ilinalta's Deep, you will notice it's partially underwater. Follow the marker arrow and use the hatch to gain entrance. Once inside, the path is pretty linear, but there are a ton of necromancers and skeletons lurking in the halls.

Be extremely careful, as some enemies will be raised from the dead by alerted necromancers. Focus on killing the necromancers as quickly as possible to keep this from happening.

You will eventually get to a room with a raised bridge — jump in the water and find the steps that lead up to the other side. Here you will find a necromancer, defeat him and pull the chain on the wall to let the bridge down before proceeding forward.

The dungeon is full of potions and enchanted items, so make sure to collect items as you progress. There is also an Alteration Skill Book behind a waterfall.

Keep proceeding through the dungeon until you reach the final room with Malyn Varen's skeleton in a chair. Directly below the skeleton will be the Azura's Star — pick it up and head back to either Nelacar or Aranea Ienith.

Bringing The Star To Aranea Or Nelacar

The outcome of the quest changes depending on who you bring the Star to. They will treat the Star based on their different motivations, leading to two outcomes.

Aranea Ienith	Gives you Aranea as a follower and Azura's Star, an unbreakable soul gem that can trap white souls.
Nelacar	Gives you The Black Star, an unbreakable soul gem that can trap white and black souls.

Black souls are captured from sentient beings such as the ten playable races and are equivalent to Grand Souls. The use of black soul gems is taboo and is even banned at the College of Winterhold. They are, however, a convenient source of potent enchanting souls.

No matter who receives the Star, you will have the final task of battling Malyn Varen and his three Dremora. If you wish to complete the task quickly, simply killing Malyn will complete the quest and allow you to exit the Star.

A DAEDRA'S BEST FRIEND

Starting A Daedra's Best Friend

The main purpose of this quest is to unite a stray dog, Barbas, with his master. You'll need to be at least level 10 to begin this quest.

Once you are, head to Falkreath. You'll find a guard asking if you've seen a seemingly-stray dog outside of town. He'll tell you to talk to Lod, a blacksmith, who will task you with catching this dog. You'll also be handed some raw meat to lure the dog with.

You can persuade Lod to give you some of the reward in advance. It's not much, but it is gold.

Head south from here to begin your search. You'll eventually find the dog that everyone is talking about, and it turns out that dog also has a lot to say. He'll introduce himself as Barbas, before asking for your help settling a quarrel he had with his master.

You can bait the name of his master out of him, and you'll learn that it's the Daedric Prince of wishes, Clavicus Vile. You'll need to head to where Clavicus Vile's shrine is located — Haemar's Shame.

Clearing Haemar's Shame

Accompany Barbas to Haemar's Shame, or, if you want, you can tell him you have other things to do and meet him there later. Should you choose to meet him there another time, he'll travel on his own, and you can save yourself some extra work.

Should you decide to accompany Barbas, you'll have to put in some extra elbow grease. The journey is long, and Barbas is easily distracted.

On this journey, you'll find that Haemar's Shame is very far from Falkreath. Along the way, you'll have to deal with any enemies you encounter, including at least two separate bandit encampments. Luckily, Barbas can't be killed, so you don't need to worry about protecting him.

Once you arrive at Haemar's Shame, you'll venture inside with Barbas to find the entire dungeon filled with leveled vampires who worship Clavicus Vile. Nothing here is too out of the ordinary.

Vampires may inflict Sanguinare Vampiris, the disease that turns you into a vampire. Be sure to check if you have it and remove it by praying at a shrine or drinking a Potion of Cure Disease once you have cleared Haemar's Shame.

You just fight your way through Haemar's Cavern and the opponents, then into the area of Haemar's Shame proper, which contains a Frostbite Spider and some other foes, before you find the shrine of Clavicus Vile.

If you need, you can find an arcane enchanter and alchemy lab along the way.

The Shrine of Clavicus Vile

Once you reach the room where Clavicus Vile's shrine is, there will be a boss to fight: a leveled vampire that Barbas will be fighting when you arrive. Dispatch him and then approach the statue and speak with it.

Clavicus Vile will begin to speak with you, telling you that he is willing to strike a deal with you. He'll then go on to express his disdain for Barbas, stating that he knows a way that you can both achieve something you want. He'll task you with retrieving his Rueful Axe from Rimerock Burrow.

It's worth noting that there's a chain behind the statue you can pull for a shortcut back into the room, so you don't have to take the long way when you return.

Retrieving The Rueful Axe

You'll need to head to Rimerock Burrow, which is located in the far northwest corner of the entire map. Barbas will accompany you, per Clavicus Vile's instruction, but you can tell him to stop following you if you wish.

Once you arrive at Rimerock Burrow, you'll find a leveled atronach inside. Head to the next room, where you'll find the mage Sebastian Lort.

Lort uses some pretty decent Conjuration spells to attack but isn't that difficult to take down. He may summon a Flame Atronach during the fight.

After this fight, grab the Rueful Axe and head back to Clavicus Vile's shrine in Haemar's Shame.

Killing Barbas Or Letting Him Live

Clavicus Vile, always the trickster, will try to convince you that you need to kill Barbas if you want to keep the Rueful Axe. Barbas will, fittingly, protest. You have to choose between two choices here.

Getting the Rueful Axe	You will have to kill Barbas with the Rueful Axe to keep it for yourself. Barbas technically does not die, though he will take several centuries to reappear, and clearly feels the pain of death.
Getting the Masque of Clavicus Vile	Telling Clavicus Vile to take back Barbas will give you the Masque of Clavicus Vile.

If you choose the Masque, you won't get to keep the axe, but you'll be able to earn the Oblivion Walker achievement and walk away knowing you didn't kill a dog.

Choosing the Axe will prevent you from earning the achievement.

Either way, you'll be done with this quest. Return to Lod in Falkreath for the small amount of gold he promised you.

Rueful Axe Vs The Masque of Clavicus Vile

Here are the stats and other info about the two items:

The Rueful Axe is not a true Daedric Artifact. This axe is the slowest melee weapon in the game, but has a damage rating of 22 and does 20 points of Stamina damage. It can be upgraded with an Ebony Ingot but does not benefit from any Smithing perks.

The Masque of Clavicus Vile is a true Daedric Artifact. It will earn you respect from anyone you meet. It will improve prices by 20 percent and improve the Speech skill by ten points. It also causes Magicka to regenerate five percent faster.

A NIGHT TO REMEMBER

Starting A Night To Remember

You start this "quest" by talking to Sam Guevenne, a black-robed stranger appearing at a tavern.

You can only start this quest at level 14. Sam will appear at the first inn you visited when you reached that level.

He'll offer to give you a staff if you win a drinking contest against him, and he'll egg you on throughout the process. Once you've downed the third one, you'll pass out and wake up in Markarth's Temple of Dibella.

The Temple

The Temple will be trashed, and a priestess named Senna will berate you for being a drunkard who made a mess of the temple. She will demand that you clean up your mess. There are more important things to do, of course. You have three options for resolving the situation.

Clean Up	If you feel some sort of responsibility and regret, you can pick up your trash. Doing so will get you a Giant's Toe, two bottles of wine, a Hagravens' Feather, and a note from Sam.
Bribe Her	Housework might not be your strong suit, and it might be better for everyone involved if you paid her to clean up instead.
Persuade Her	You can plead with Senna to let you off, provided that you are a smooth talker. If you succeed, you will be let off.

Regardless of whether you clean up, persuade Senna, or bribe her, she'll tell you that when you came in, you had been blabbering about Rorikstead.

Rorikstead

Make your way to Rorikstead, and you'll be confronted by a Redguard named Ennis, who says you took his goat, Gleda, and sold her to a giant named Grok. He'll basically tell you that you need to retrieve Gleda, or else. Like with Senna, you can weasel your way out.

Help Him Get Gleda	You will have to get Gleda away from the giant. Once you get close enough, interact with her to get her to follow you. Grok will try to stop you.
Bribe Him	You can pay him for his goat. It's not like they're hard to find, anyway.

Persuade Him	A simple lie will convince him that you need to find Sam to get his goat. You do not have to return.
Intimidate Him	He's just a goat farmer anyway. Flexing your muscles and giving a threat will get him to back down.

Should you choose to retrieve Gleda, you'll need to be careful. Grok is located south of where you'll find Ennis, with Gleda nearby. It is possible to sneak up to Gleda, "talk" to her, and get her to follow you without Grok noticing, but this can be difficult.

Grok will not be hostile at first, but will attack when you try to take Gleda. Additionally, your follower may attack him when you're trying to be sneaky. Try casting Calm on Grok if you have a decent Illusion level and the right perks.

Pass any of those challenges or retrieve the goat and Ennis will tell you that you left a note with the phrase "after repaying Ysolda in Whiterun."

Repaying Ysolda

Apparently, you owe Ysolda money for a wedding ring, which may be confusing to you. The wedding was called off though Ysolda wants the ring back now. Once again, you have to decide how to resolve the situation.

Get The Ring	You will go to Witchmist Grove to get the ring.
Pay Her	The ring costs a hefty 2,000 gold. She will be satisfied if you pay it off.
Persuade Her	With the power of love, you can compel her to tell you where to go.
Intimidate Her	She is tougher than she looks and is not easy to intimidate. She will back down if you are menacing enough, of course.

If you succeed at the persuasion, bribe, or intimidation, you can head straight to Morvunskar, where Ysolda tells you the wedding was supposed to be taking place.

❖ Going To Witchmist Grove

However, if you should decide to retrieve the ring, you'll need to head to Witchmist

Grove. Your bride-to-be is a Hargraven named Moira. Ask for the ring back, and she'll assume you're going to leave her for another woman named Esmerelda, then she'll attack you.

After the fight, grab the ring and return it to Ysolda, who will tell you that you were meant to be attending the wedding at Morvunskar.

The "Wedding"

No matter how you learn that the ceremony is meant to take place at Morvunskar, head there next. Inside this ruined fort is a few hostile mages that you'll need to deal with. Eventually, you'll find a glowing orb that transports you to Misty Grove, where you'll finally find Sam, standing around a table with a few guests.

It turns out that Sam is actually the Daedric Prince Sanguine, the deity of Debauchery. In exchange for raising hell throughout Skyrim, he gives you the staff, the Sanguine Rose, a Daedric Artifact.

This staff summons a Demora to fight for you for sixty seconds. You can give followers the Sanguine Rose, and they'll be able to successfully use it as well.

After you receive the staff, Sanguine will teleport you back to the tavern where you first met him, ending the quest.

ILL MET BY MOONLIGHT

Talk To Sinding

To begin Ill Met By Moonlight, you don't need to reach a certain level cap, unlike the majority of other Daedric quests, and can instigate it whenever you want. However, since this quest involves fighting higher level enemies, it might be best to wait until you are a higher level yourself. Since the quest rewards you with either the Ring of Hircine or the Savior's Hide, it might be best to complete the Companions questline first and become a werewolf yourself, as this will help in the final conflict. After talking to Sinding, he will give your the Cursed Ring of Hircine, which is automatically equipped and can't be unequipped, before escaping prison and fleeing to Bloated Man's Grotto.

Get Out Of Falkreath ASAP

Unfortunately, wearing the Cursed Ring of Hircine will cause you to transform into a werewolf at random, even if you weren't a werewolf when you started the quest. With this in mind, run out of Falkreith as fast as you can to avoid accidentally transforming in front of the Hold guards and instigating a fight.

Once you have successfully fled Falkreath, follow the quest marker to find the White Stag and begin the next phase of the quest. En route, you may have to deal with a handful of bandits, including Bandit Chiefs, so make sure you are always prepared for a fight at any given moment.

Find The White Stag

Like all stags in Skyrim, the White Stag is very hard to hit with melee weapons and should be killed using a bow or magic, as getting close to the creature without scaring it off is next to impossible without a very high Sneak skill level. After killing the stag, the Aspect of Hircine will appear and commend your hunting capabilities, taking note of the fact that you are wearing the Ring of Hircine. Whether you accept Hircine's request or not, they will send you to Bloated Man's Grotto, where Sinding is hiding, so that you can end the fight once and for all.

Track Down The Wild Hunt

After a brief conversation with Hircine, you will enter Bloated Man's Grotto and find that Sinding is single-handedly dealing with the Wild Hunt in his werewolf form. After talking to a dying Khajiit hunter, make your way into the main dungeon area to find Sinding, who pleads with you to help him kill off the rest of the Wild Hunt and leave him in peace to live in the Grotto. If you accept his offer to help, you can simply follow him from a distance while he tears through the Wild Hunt, otherwise, you will have to battle him immediately.

Spare Sinding

If you chose to spare Sinding, you will need to fight off multiple low-level enemies that are relatively easy to deal with, especially in werewolf form. While the majority of the Wild Hunter is wearing fairly weak armor and can be defeated pretty easily, there are a couple that are wearing Steel Armor and can be much harder to defeat at low levels.

After fighting off the Wild Hunt, Sinding will thank you for your help and the Cursed Ring of Hircine will simply become the Ring of Hircine and can now be unequipped as well as no longer randomly triggering werewolf transformations.

Talk To Hircine

After helping Sinding and leaving the Grotto, the Aspect of Hircine will appear to you as the stag once again, standing at the entrance to confront you, where you can either say you failed Hircine or continue to defy them. If you take the dialogue option where you admit you failed to kill Sinding, Hircine will give you another chance to rectify this error and send you back inside the Grotto to finish the job. If, instead, you chose to kill Sinding, he will appear to you as Sinding beside his corpse and congratulate you on a job well done, before turning the skin into the Savior's Hide.

Tear The Skin From Sinding

If you refuse to help Sinding, or if Hircine sends you back to finish the job, you will have to battle against a fairly difficult werewolf, especially if you are at a low level. If you can transform into a werewolf, even at low levels, this battle should be fairly easy, however, and you should be able to knock him around with ease. After killing him, interact with his corpse to skin Sinding, which will trigger Hircine to appear once again. They will commend your good work and turn Sinding's skin into the Savior's Hide, a low-level armor piece that lacks in armor rating but has decent enchantments with Magicka and Poison Resistance.

It is possible to obtain both the Ring of Hircine and the Savior's Hide as quest rewards, despite the fact that the quest is designed to only give you one of them. If you run out of the Grotto fast enough, you should be able to catch Hircine in time for them to give you a second chance once the Cursed Ring of Hircine has been altered, though you should have before attempting this as it doesn't always work and can result in you having skinned Sinding for no reason.

BREAK OF DAWN

Starting Break Of Dawn

There are two ways to start this quest. The first is to play Skyrim normally and clear out dungeons. Once you reach level ten, the loot at the end of a dungeon may contain Meridia's Beacon.

Picking it up will have the Daedric Prince of Life bombard you with a demand to cleanse her temple of a necromancer. As a being that despises the undead, she will not suffer the insult.

If you had thought that the Beacon was clutter and left it in a chest, visiting her Shrine will give you a quest marker to the object, allowing you to continue with the quest.

The second way would be to visit her shrine located by the road between Solitude and Dragon Bridge. Speaking to the statue will start the quest and point you to a dungeon where you find the Beacon.

She gives you the next instructions when you return the Beacon to her statue. You will have to guide a beam of light through the desecrated temple to restore it to glory.

Walk down the hill to reach the entrance to Kilkreath Temple.

Clearing Kilkreath Temple

Kilkreath Temple is filled with the desecrated corpses of the necromancer's victims and their hostile shades. You will need to fight them as you raise pedestals throughout the temple to clear the way ahead.

The desecrated corpses often contain hefty sums of gold, making pilfering them a lucrative activity.

The first room familiarizes you with this mechanic. Walk to the pedestal and interact with it. Doing so will reflect the beam to the next beacon and open a door on the south side of the room.

In case you are curious, walking through the beams will deal Shock damage over time and very quickly.

Go through that room to find a short hallway with a handful of shades. At the very end, you will reach another room with a single pedestal at the center. Interact with the pedestal to open a door to Kilkreath Balcony.

The Balcony itself lacks enemies, though watching the beacons will show you the light spreading through the temple. There is also a large, locked chest by the final set of stairs that contains leveled loot.

Solving Kilkreath Ruins

Kilkreath Ruins is where the temple gets a little confusing. Start by clearing out the Shades spread throughout the room. Go to the east pedestal, up a set of stairs to open a doorway next to it. This brings you to a long hallway with a few more Shades.

Keep your eyes peeled for an axe trap in the middle of the hallway.

At the very end, you will be able to reach the next pedestal that is isolated and without any connections to adjacent platforms. While it may look impassable, sprinting across will get you to the pedestal. Interact with it to open the path to the next pedestal. There will be no need to leap across; it is connected to the rest of the room by a doorway on the west.

Finally, go through the door and fight another Shade. Go up a set of stairs to the east to reach the final pedestal. Interacting with it will open the path to Kilkreath Catacombs.

Defeating Malkoran At The Kilkreath Catacombs

The final part of the quest needs you to defeat the necromancer Malkoran, the one responsible for desecrating the temple. Not only is he surrounded by several Shades, but he is also capable of casting powerful Frost magic and is not afraid to use it.

Malkoran will stay at the altar throughout the fight to sling magic from relative safety, thanks to the Shades separating you. If you can keep the Shades distracted, it is possible to snipe Malkoran before continuing.

Once defeated, Malkoran will transform into a Shade and immediately resume attacking you. Be prepared for this and top up your health prior to defeating him. Vanquish him a second time to satisfy Meridia.

Getting Dawnbreaker

At the very end, Meridia will ask you to pick up Dawnbreaker from the altar. Once again, she lifts you to the sky, where she declares you her champion. Regardless of how you answer, she returns you to the ground with a new weapon in hand.

Dawnbreaker is a powerful one-handed sword that excels against the undead. Its unique enchantment deals ten points of fire damage to the target. If an undead foe was defeated using its enchantment, there is a chance for it to explode. Nearby undead will be set alight and may suffer the effects of Turn Undead.

As vampires are considered undead, the explosion will damage and possibly kill you. It will also affect undead allies such as Serana.

The weapon can be improved with ebony ingots at a grindstone with the Arcane Blacksmith perk.

THE TASTE OF DEATH

Starting Taste Of Death

To begin The Taste of Death, head inside Markarth, where the city's Hall of The Dead has been closed to the public. To gather information about the Hall, players can choose to gather information around town or go to Brother Versulus. Brother Versulus can be found outside the entrance of the Hall and will ask you to investigate the area.

Upon entering the Hall, you will hear a voice. Venture a little further inside to meet Eola, who will invite you to Reachcliff Cave to help her with a Draugr problem before disappearing.

It's important that you do not attack Eola as it will immediately fail the quest.

To make things a bit easier, later on, go back to Brother Versulus and let him know that the Hall of The Dead is clear. He will then start staying in the Hall throughout the day.

Dealing With Reachcliff Cave

This is another simple part of the quest; you simply need to venture to Reachcliff Cave and speak with Eola again. She will task you with wiping out the Draugr inside, which can be easily done at just about any level. You will simply need to process to the end of the cave, ensuring that you have defeated all the enemies.

The final room is quite distinct, as there is a giant table in the middle. Clearing the cave will prompt Eola to appear and ask you to lure Versulus to the cave to become the cannibal's next meal.

Completing The Feast

ATo find Versulus, simply go back to the Hall of The Dead, and give him the bribe that Eola presented to you. It's important that you do not have a follower with you at the time, as Versulus will need to fill the follower slot.

Versulus will then follow you to the cave, where he will be persuaded by Eola to lie down on the sacrifice table. To complete the quest, simply hit Versulus to kill him and then choose to feast on his flesh by interacting with his body. This will award you with Namira's Ring.

The ring will give you an extra 50 stamina. It also gives you the ability to feed on corpses.

Each feeding session will replenish 50 health and boost health regeneration by 50 percent for five minutes.

THE HOUSE OF HORRORS

Starting House Of Horrors

To begin this quest, head to the Silver-Blood Inn in Markarth and speak with the innkeeper, who will tell you that a member of the Vigilants of Stendarr has been spotted in town. This person is Vigilant Tyranus, a witch hunter. Supposedly, he's been pestering the locals with questions, specifically about an abandoned house in the city.

Alternatively, you might see Tyranus outside the Abandoned House talking to a local under the right conditions.

As a Vigilant of Stendarr, his order will be diminished after you start the quest Dawnguard.

Tyranus will bombard you with questions too, and unless you completely shut him down, he'll introduce himself and ask for your assistance. Agree, and he'll bring you to the abandoned house.

Inside The Abandoned House

Once inside, Tyranus will be a bit perplexed, as he notes that there are no signs of the house aging, and fresh food is sitting out. Shortly after this, a loud noise will spur him on, and he will try to investigate its source. Unfortunately, there will be a locked door in the way.

Once you interact with the door, a voice will begin speaking to you, and the furniture will begin floating. Tyranus will be frightened and run away, just to find you are both locked inside the house.

Again, interact with this door to hear the voice speak to you again. It'll continue telling you to murder Tyranus, now threatening that if you don't, it might be your life on the line instead.

However, it quickly becomes apparent that Tyranus is hearing the same voice — and it's telling him to do the same thing to you. He'll succumb to the voice and attack you in hopes of keeping himself alive.

You will not be able to continue the quest unless Tyranus is killed. He uses some spells, like sparks, and defensive Alteration spells. He is not particularly dangerous.

Win the fight, and the voice will tell you that you have earned a reward, so head into the tunnel where you're directed.

Meeting Molag Bal

On the other side of this tunnel, you'll find an altar with a rusty mace atop it. Once you try to claim it, you'll be caged. The voice, it seems, is Molag Bal, the Lord of Domination. His main priority as a Daedric Prince is harvesting mortal souls. He'll ask you what you see.

Choose any of the three dialogue options, and he'll tell you that the mace was once powerful, but it and the altar on which it stands were desecrated by Boethiah's priests.

Molag Bal will then explain (after the next dialogue prompt) that the priest of Boethiah typically comes to desecrate the altar but hasn't shown up in a while. He's imprisoned somewhere else. Molag Bal wants you to find him and bring him back.

Finding Logrolf The Willful

This priest is named Logrolf the Willful, and he'll be imprisoned in one of six random locations, all of which are in the Reach and filled with Forsworn.

The location where he'll be found will be marked on your map, but can be any of the following: Broken Tower Redoubt, Bruca's Leap Redoubt, Deepwood Redoubt, Druadach Redoubt, Hag Rock Redoubt, or Red Eagle Redoubt.

Once you find him, you'll want to make sure you've dealt with all the Forsworn, because they'll attack Logrolf as makes his way out. He'll be a bit hesitant to believe that you just came to rescue him out of the pure goodness of your heart.

You'll have to either persuade, bribe, or intimidate him. You'll need a Speech skill of at least 50 to persuade him successfully. Once you untie him, he'll run off to Markarth and the abandoned house.

The quest will fail if Logrof dies, so be sure there are no Forsworn in his way and that you don't harm him yourself.

Getting The Mace Of Molag Bal

Once back in the house, Molag Ball will imprison Logrolf. A conversation will begin ultimately leading to Molag Bal telling you to do something horrible. You'll need to attack Logrolf. You'll find the rusty mace is equipped, and you need to use it to beat him repeatedly. Every time he dies, Molag Bal will revive him, and you'll have to do it all over again.

Eventually, Logrolf will submit and revoke his pledge to Boethiah. The Daedric Prince will task you with finishing Logrolf off one final time, and once you do, he'll turn the rusty mace into his signature Daedric Artifact. This is the Mace of Molag Bal. The quest will then finish. Whether it was worth it is up to you... and your conscience, of course.

❖ Is The Mace Of Molag Bal Good?

Also known as the Vampire's Mace, the Mace of Molag Bal is a Daedric Artifact.

Every time you strike an opponent with it, they'll be drained of Magicka and stamina (25 points each). It is also enchanted with soul trap, inflicting it on the target every three seconds.

If you want to upgrade it, you'll need an ebony ingot and the Arcane Blacksmith perk.

Like other unique items, the Mace of Molag Bal does not become more powerful as you level up. It will become increasingly obsolete as you gain experience and strength.

If you walk around Skyrim with the mace equipped, bystanders will comment on it, asking you to get it away from them.

THE ONLY CURE

Talk To Kesh

To begin the quest, you need to reach a minimum of level 10 to find Kesh at the Shrine to Peryite, located the dragon shrine northeast of Markarth. The quest can also be triggered by encountering a radiant Afflicted NPC, which will direct you to find Kesh as well. After talking to Kesh the Clean, you will be directed to find certain ingredients for Kesh: a Deathbell Flower, a Silver Ingot, Vampire Dust, and a Flawless Ruby. While the majority of these items are relatively easy to find, the Flawless Ruby is much more difficult and can be the sole reason why this quest can take much longer to complete than it otherwise should.

Find The Necessary Ingredients

Thankfully, only one of the ingredients needed is hard to find, while the others can all be found relatively easily. Deathbell Flowers can be obtained by delivering Farengar's Frost Salts to Arcadia in Whiterun, while being found in the wild around Morthal fairly consistently. Vampire Dust can be obtained by defeating any vampire enemy, though can also be bought from several Alchemy vendors, including Arcadia in Whiterun. Silver Ingots are a little harder to track down, though most general vendors or Smiths will sell them from time to time. You can also find the Transmute Mineral Ore spell in Halted Steam Camp, just behind Whiterun, that can turn Iron Ore into Silver Ore when cast.

Flawless Rubies, unlike regular Rubies, are very hard to find at low levels. You can increase your chances of finding a Flawless Ruby in chests by finishing the No

Stone Unturned quest, which grants the player a passive buff that increases the chance to find any gemstones in chests. However, this quest is easily one of the longest and most difficult to complete in Skyrim, especially without a guide.

Although you can find a Flawless Ruby randomly throughout the game, one of the easiest ways to obtain one is to complete the Dark Brotherhood questline, as one is always found in the Emperor's Quarters on the Katariah, while another can be looted from Amaund Mottiere's corpse. You can also attempt to reset for one by completing a short errand for Harrald Law-Giver in Riften, where he gives you two randomized gems for completing his errand. While you can get a Flawless Ruby from this errand, Flawless gems appear much less frequently than regular variations and may take multiple resets before one is given.

Kill Orchendor

After returning to Kesh with the necessary ingredients, Peryite will appear and claim that his champion, Orchendor, has failed him and that you are needed to kill and replace him as Peryite's Champion. Once the dialogue with the Daedric Prince has ended, you will need to travel to Dwemer ruin Bthardamz, where Orchendor is bunkered down with dozens of Afflicted. This ruin can only be entered through this quest and you won't be able to get past the main door without first talking with Peryite.

After finding Orchendor, deep inside the dungeon, it becomes apparent fairly quickly that he is no pushover. Aside from being entirely immune to all spells, he is also able to teleport around the battlefield to avoid melee attacks and will be joined by several Dwemer automatons, including a Centurion.

The best way to approach this battle is from a distance with a bow or crossbow since you should be able to land at least one sneak attack on him before he decides to teleport around the battlefield. This way, you also avoid being flanked by Dwemer automatons while trying to fend him off, as they can deal quite a lot of damage if left unchecked. Once Orchendor is defeated, loot his corpse and take the key to the Bthardamz elevator, which will prevent any longwinded backtracking, and return to Peryite for your reward.

Return To Peryite

Upon returning to the Shrine to Peryite, inhale the fumes once again to talk to the Daedric Prince, where you will have to have a short conversation with them congratulating your ability to kill people who were well on their way to a long and painful death before the quest is completed. Once the dialog is over, you will be given Spellbreaker as a reward, a shield with a base armor rating of 38, even higher than a Daedric or Dragonplate Shield. Once you have this shield, there is almost no reason to use any other shield in the game, with the only exception being if you want to use a shield with a unique effect, such as the Targe of the Blooded or Auriel's Shield.

THE MIND OF MADNESS

Starting The Mind Of Madness

You start your journey by looking for a raving beggar called Dervenin near the Bard's College in Solitude. He will tell you that his master is on a vacation and that there will be disastrous consequences if you do not help him resolve the issue.

Dervenin's master is in the Pelagius Wing within the Blue Palace. The beggar will hand you a hip bone to allow you to find his master within. It is best not to think too hard about it.

Accessing The Pelagius Wing

The Pelagius Wing is located inside the Blue Palace, in the Avenues District. Once inside, you will quickly learn that getting a key to the Pelagius Wing doesn't seem to come easy. Surprisingly though, there aren't any extra hoops you need to get through to get into the room. To get the key, simply talk to one of the maids (like Erdi) to get the key for the wing.

You could also get the key from Falk if you finished The Man Who Cried Wolf.

Once you proceed inside, you will find yourself in an abandoned, cobweb-covered part of the Blue Palace. Simply walk forward until you find yourself transported inside the mind of Pelagius the Mad, an infamously murderous Emperor.

Escaping Pelagius' Mind

Here you will find Sheogorath, the Daedric Prince of Madness talking to Pelagius in a demented therapy session. He refuses to leave until you "cure" Pelagius. He will hand you the Wabbajack and send you to treat the dead Emperor's personal demons by solving three different puzzles.

The quest is now divided into three separate parts that can be done in any order.

❖ Anger Issues

Anger Issues is accessed by using the southeast portal. There you will find Confidence, a tiny rendition of Pelagius, being pummeled by Anger, a large Imperial soldier. You need to use the Wabbajack on Confidence to make him grow.

Growing Confidence will create Self Doubts that will beat Confidence to a smaller size. Be mindful of their appearance and plan your adjustments accordingly.

Your target is to grow him large enough to take out Anger. Simply use the item on Confidence to make him grow and on Anger to make him shrink. Once you hear Sheogorath begin speaking, the task is done.

❖ Paranoia

Next, you can access Paranoia by using the northwest portal. Here you will find an arena where Pelagius is fighting two Storm Atronachs along with a couple of bodyguards. Here you need to ignore the Atronachs and shoot the bodyguards to turn them into wolves.

❖ Night Terrors

Finally, you will need to use the northeast portal to deal with Pelagius's Night Terrors. This is where things can get a bit tricky, as you will need to jump back and forth between shooting different enemies. Here is the order the Wabbajack needs to be used in:

1. Shoot sleeping Pelagius
2. Shoot the wolf
3. Shoot Sleeping Pelagius
4. Shoot the Bandit
5. Shoot Sleeping Pelagius
6. Shoot the Hag
7. Shoot Sleeping Pelagius
8. Shoot the Flame Atronach
9. Shoot Sleeping Pelagius
10. Shoot the Dragon Priest

Once the Priest turns into a chest and Pelagius wakes up, you have completed this task. Return to Sheogorath to report your successful therapy. He will be confronted by Derevin and will be convinced to end his vacation.

Getting The Wabbajack

Before he leaves, Sheogorath allows you to keep the Wabbajack as your reward. You will be able to use the staff as a weapon in combat.

The Wabbajack has over two dozen effects that will be randomly applied to a victim It is not the most powerful weapon, but it is certainly very entertaining.

In a perfect representation of its creator's domain, the Wabbajack will either help or harm you when used with no way of predicting the outcome. It could unleash an offensive spell, transform the target into livestock, or simply kill the target outright.

WAKING NIGHTMARE

How To Start The Quest

Starting the "Waking Nightmare" quest is rather simple due to the lack of prerequisites. Travel to Dawnstar and speak to a priest named Erandur in Windpeak Inn. He will talk about the town's nightmare curse, theorizing that Vaermina is behind it. Erandur will ask you to venture to Nightcaller Temple, a tower looming over the Hold.

Accepting his offer will start the quest. Escort him to the tower over the nearby hill, taking down any frost trolls or other wildlife that inhabit the path. After some exposition about the origins of the curse, head inside the temple to start the quest's first step.

Explore Nightcaller Temple

The first part of this quest involves following Erandur through the temple. Stick by his side as he leads you to the source of the nightmares. Upon reaching an energy barrier further down the temple, Erandur will reveal himself to be a priest of Vaermina that fled the temple long ago.

When a band of Orcs arrived at Nightcaller Temple to kill its inhabitants, Erandur left and became a priest of Mara. He hopes that he can right the wrongs of this temple by destroying the source of Dawnstar's nightmares: the Skull of Corruption.

Continue to follow Erandur until he leads you to a library, killing any awakened Vaermina priests as you go. To remove the barrier, you will need to find The Dreamstride book and Vaermina's Torpor.

The library will have the book, but you'll need to explore more of the temple to find the potion. You can find the book on the top floor on a pedestal. Bring the book to Erandur to start the next step.

Find Vaermina's Torpor And Drink It

Erandur will describe the potion as a small flask. In-game, it has the appearance of a Potion of Ultimate Healing with a darker shade of red. Vaermina's Torpor rests in the laboratory next to the library you found The Dreamstride book.

It will be residing on a shelf on the laboratory's bottom floor beside a whole host of alchemy ingredients.

Upon bringing the potion to Erandur, he will ask you to drink it. Consume the item in your inventory to start a strange flashback sequence where you take control of a Vaermina priest. Veren Duleri will ask you, a priest named Casimir, to release the Miasma on the orc invaders.

To do so, run past the fighting enemies until you reach a statue of Vaermina. Pulling a chain in front of the statue will release the Miasma.

It turns out that the barrier blocking further exploration was at this statue. You will abruptly be pulled back to the present in this same spot. To remove the barrier, grab the soul gem beside the Vaermina statue. This will allow you to regroup and Erundar and open the path to the Skull of Corruption. Trek down to the bottom of Nightcaller Temple to deal with the Daedric Artifact.

Side With Erandur Or Vaermina

Both you and Erandur must fight Vaermina priests Veren and Thorek upon reaching the Skull of Corruption. In a not-so-shocking twist of events, Erandur is revealed to be Casimir. Kill both Veren and Thorek.

Once done, Erandur will prepare a spell to destroy the Skull of Corruption once and for all. Vaermina begins to talk to you once Erandur starts the ritual, stating that he's deceiving you.

You have two options at this point:

1. Kill Erandur and take the Skull of Corruption.

2. Wait until Erandur finishes the ritual, destroying the Skull of Corruption for good.

Attacking Erandur at this point will make him hostile, forcing you to kill him. Fortunately, he's a weak opponent. Either best him or let Erandur complete the ritual to finish the quest.

Note that if you're interested in earning the Oblivion Walker achievement, you must kill Erandur. If Erandur destroys the Skull of Corruption, it voids this achievement for the rest of your playthrough.

Waking Nightmare's Quest Rewards

Completing this quest will grant either Erandur as a potential companion or the Skull of Corruption Daedric Artifact.

❖ Siding With Erandur

Completing this quest on Erandur's side will unlock him as a companion. Erandur is a Dunmer skilled in all magic schools, including Restoration and Conjuration. He prefers to use Destruction spells such as Firebolt and Flames, although he is more than capable of using a mace or sword. He makes for a solid companion for archer or mage builds that prefer to keep their distance.

❖ Siding With Vaermina

Killing Erandur will grant the Skull of Corruption as your quest reward. This staff deals 20 points of magic damage in a large area. Uniquely, using this staff on sleeping NPCs will charge the staff. This increases its damage output to a cap of 50 points of magic damage.

Using the Skull of Corruption on sleeping NPCs will not make them hostile. Only five dreams may be harvested from any given NPC.

Mages will get more mileage out of using Destruction spells. That said, builds that prefer to use melee weapons can get some excellent mileage out of this staff. Just be careful using the Skull of Corruption in enclosed spaces; the AoE damage can harm allies.

BOETHIAH'S CALLING

How To Start The Quest

To start Boethiah's Calling, you either need to find and read Boethiah's Proving or find the Sacellum of Boethiah. Boethiah's Proving spawns on dead Boethiah followers and in certain dungeons.

You will not be able to start the quest until you are level 30. The books and cultists will not spawn, and nothing will happen at the shrine until you reach that level.

Suppose you want to skip the book and find the shrine yourself, head east of Kynesgrove until you reach Traitor's Post. Take the right fork on the nearby road to reach the shrine.

Kill In Boethiah's Name

Cultists will surround Boethiah's statue when you approach. Talk to the priestess to start the first quest objective. She will inform you that Boethiah demands a sacrifice. The priest will give you the Blade of Sacrifice and ask you to kill someone you trust.

There are two ways to proceed:

- Lure a non-essential follower to the Sacellum of Boethiah. Order them to interact with the Pillar of Sacrifice, then kill them.
- Kill every Boethiah cultist at the shrine.

❖ Sacrificing A Friend

You will need to find a willing follower and bring them to the Sacellum of Boethiah. Only non-essential followers will work for this. Essential followers, such as Aela the Huntress or Frea cannot die and will not be suitable for this step.

If you don't wish to kill a Housecarl or useful companion, you can find a hireling for the sacrifice.

If 500 gold is too expensive, head to Riverwood and help either Faendal or Sven. They give you a short quest that lets you take one as a follower. Their lives are free and nothing of value will be lost by sacrificing them.

At the Sacellum, hold down the interact key and direct your hapless victim to the

Pillar of Sacrifice. Once they are trapped, slay them.

❖ Killing The Cultists

Killing every cultist at the Sacellum of Boethiah also works. These cultists can put up quite the fight if you're under-geared, so be sure to bring upgraded weapons and armor before attempting this. Area of effect spells and Shouts do wonders here.

Speak To Boethiah

Whether you killed every cultist or sacrificed a follower, Boethiah will begin to speak through a corpse. If you killed a companion, Boethiah will state that she has a task for whoever is left standing. This will cause all cultists to become hostile and attack each other. Kill the cultists to proceed.

You can hang back and let the cultists kill each other instead of getting involved and potentially dying.

Once the cultists are dealt with, Boethiah will offer to make you her champion under one condition: her old champion must die.

Kill The Bandits In Knifepoint Ridge

Boethiah will direct you to Knifepoint Ridge, a fortified bandit camp northwest of Falkreath. She wants you to kill them stealthily. Should you kill her champion and take his Ebony Mail, Boethiah will crown you as her new champion. Travel to Knifepoint Ridge to take the bandits out.

Despite what Boethiah requested, you do not need to use stealth in this dungeon. She will chastise you for using brute force, but this has no impact on the quest. There are two sections to Knifepoint: an exterior bandit camp and an interior mine Boethiah's Champion can be found in the mine.

Most of this dungeon is linear. Use the shadows, flammable traps, or use sheer aggression to take down the bandits. Once you reach the lowest section of the mine you will enter a room with the Champion of Boethiah, along with a few bandit guards.

If you can calm the champion, he will have a few lines of dialogue explaining his change in attitude. In essence, he became tired of killing for the sake of a Daedra's entertainment. He is content with dying by your hand.

Killing the champion up-close is somewhat challenging due to the Ebony Mail's enchantment. Using ranged weapons or spells to damage the champion is a good idea. Once you take his life, equip the Ebony Mail chest plate on his corpse to end the quest.

Getting The Ebony Mail

Regardless of how you complete this quest, Boethiah will always give you the

Ebony Mail as a quest reward. This Daedric Artifact is a heavy armor chest plate that acts as a stealth item.

Crouching with this chest plate with cloak your character in shadow. The chest plate also has the muffle enchantment, the only non-boot item in Skyrim to have this effect. Additionally, any foe that gets near you takes five points of poison damage per second.

Enemies that are usually immune to poison will take damage from this effect. This includes vampires, Dwemer automatons, and even draugr.

The armor does come with a few quirks:

1. Ebony Mail's poison enchantment will escalate brawls, turning every nearby NPC hostile.

2. The poison aura works while underwater, allowing you to kill fish while swimming.

3. Wearing additional pieces of Ebony gear makes the Ebony Mail's smoke effect stronger. This effect can obscure your first-person view.

4. The armor's smoke effect only activates when nearby enemies become hostile.

Ebony Mail is best suited for ranged stealth characters. Melee stealth characters will get frustrated with the poison damage constantly alerting enemies. If you have a high Sneak skill, this stops becoming an issue. For ranged characters, the poison damage can be convenient when multiple enemies close the gap.

Ultimately, this item excels at giving players enhanced stealth mechanics while buffing their armor rating. If you love heavy armor and stealth, the Ebony Mail is an item you'll want to own.

PIECES OF THE PAST

How To Start Pieces Of The Past

After you reach level 20, a courier will deliver an invitation to visit a new museum in Dawnstar. Should you reach travel to Dawnstar before the courier delivers the invitation, you can walk up to the museum to start the quest.

Visit The Museum

The Museum of Mythic Dawn, known as Silus Vesuius's house in-game, is the museum the pamphlet referred to. It can be found near the coastline on the north end of town. Walking near the museum will trigger a scripted dialogue sequence between Silus, the museum owner, and court wizard Madena. Speak to Silus to get a tour inside.

Silus will allow you to browse the contents of the museum, although you can continue the quest by talking to him straight away. Oblivion fans will recognize most of the items Silus has collected. Whenever you're ready, speak to Silus once again to hear about a job offer.

He will tell you that Skyrim has fragments of Mehrunes' Razor, a powerful Daedric Artifact that belongs to Mehrunes Dagon. Silus will ask you to retrieve three pieces: the pommel, blade shards, and the hilt.

Find The Pommel

Dead Rock Crane houses the first piece of Mehrunes' Razor. Head southwest of Markarth to find this area.

Reaching this zone can be quite a pain. Starting from the Left Hand Mine outside of Markarth, head southeast across the bridge. Shortly after crossing this bridge, you will find a fork in the road. Take the right dirt path. Follow this path until you reach a bridge crossing a waterfall. Instead of crossing the bridge, turn right and climb up the hill. The top of the hill will reveal a clear path to Hag Rock Redoubt. Enter the dungeon and clear it out.

The end of the dungeon will lead to a staircase that reveals the entrance to Dead Rock Crane. Enter the tower and climb to the top. Exiting the tower will reveal a sacrificial shrine in use by a Hagraven named Drascua. Her corpse will have the pommel for Mehrunes' Razor.

Find The Blade Fragments

Mehrunes' Razor's blade fragments can be found in Cracked Tusk Keep, a fort directly west of Falkreath. Slay the bandits surrounding the camp and head inside. An Orc named Ghunzul has a key to the chest containing the shards. He can be found inside the keep on the third floor. You can either kill him or steal the key on the dresser near his bed; the result is the same.

Now that you have the key, proceed to the lowest floor of the keep. The key will unlock a button that opens up the Cracked Tusk Keep Vault. Keep an eye out for tripwires on the floor, as this area is riddled with traps. At the end of the corridor are the blade shards.

Find The Hilt

Obtaining the hilt is the easiest part of the quest. Travel to Morthal and head to the northern end of town. A man named Jorgen works at a mill for most of the day. Confront him about the hilt. He won't give the hilt to you, leaving you with four options:

1. Persuade him to give you the hilt.

2. Bribe him (this costs around 300 Gold).

3. Intimidate him.

4. Steal his key. You can either pickpocket or kill Jorgen.

Any of those four outcomes will give you the key. Head into his house to unlock the chest containing the hilt.

Meet At The Statue

With all three pieces recovered, travel back to Silus' museum to deliver the fragments. He will pay you a handsome sum of Gold. If you brought all three pieces at once, Silus gives much more Gold. The quest isn't done yet, however.

Now that Silus has every part of Mehrunes' Razor, he wants to reconstruct the Daedric Artifact. Regardless of dialogue choice, you will need to meet him at the Shrine of Mehrunes Dagon. The shrine is southwest of Dawnstar and a short walk from The Lord Stone.

Decide Silus' Fate

Asking a Daedric Prince to assist a mortal goes as well as you'd expect. Mehrunes Dagon would rather spill Silus' blood than rebuild the dagger for him. Dagon will speak to you, asking you to kill Silus yourself. Silus will then speak to you, pleading for his life. You have two choices:

1. Spare Silus, angering Mehrunes Dagon.
2. Kill Silus, pleasing Mehrunes Dagon.

❖ Spare Silus

Should you spare Silus, he will run back to Dawnstar after giving you some gold. Consequently, Dagon will spawn two Dremora enemies to kill you and Silus. Kill them or flee to end the quest. Since you don't get Mehrunes' Razor this way, this voids the Oblivion Walker achievement.

❖ Kill Silus

If you leave dialogue with Sirus or accept Dagon's request, Silus will immediately turn hostile. He is an archetypal mage, using Destruction spells to take you down from a distance. Close the gap to take him down. Interact with the shrine once he's dead to claim Mehrunes Razor. Dagon will spawn two Dremoras as one final test.

Quest Rewards

Helping Silus throughout the whole quest will give you a healthy sum of Gold. Unfortunately, this path does not grant Mehrunes' Razor—voiding the Oblivion Walker achievement.

Fulfilling Dagon's request grants his signature dagger as a reward. This weapon is effectively a Daedric Dagger while weighing half as much. Unique to this weapon, Mehrunes' Razor has a 2% chance to kill someone instantly. This effect will kill any non-essential NPC including Legendary Dragons, Giants, and Dragon Priests. Sadly, a bug prevents this weapon from benefiting from any Smithing perks. The

Unofficial Skyrim Patch fixes this bug.

ANCIENT ARTIFACTS

FORBIDDEN LEGEND

How To Start Forbidden Legends

The quest will automatically start when you learn about the three brothers. Unlike most quests in the game, you will have to stumble across clues to start the quest instead of speaking to a quest giver. Or, you could search for it directly using the methods below.

Investigating the campsite by Folgunthur	You will find an abandoned campsite by the ruins near Morthal. The quest will start when you read a journal left in one of the tents.
Defeating any of the brothers	You can pick up a Writ of Sealing from any of the defeated brothers. Reading it will start the quest.
Reading Lost Legends	This book can be found across Skyrim but is most easily located in Farengar Secret-Fire's office on a bookshelf. Reading it will start the quest.

You will be sent to Folgunthur to begin your search. If you had not started it there, you will read Daynas Valen's journal before being prompted to enter the ruins itself

Exploring Folgunthur

The first thing that you see inside the ruins are puddles of blood spilled from dead adventurers surrounded by draugr. Walk through the dungeon, following a linear path that will lead you into an open room with waiting draugr.

There is a large puddle of oil in this room. Be sure not to be caught in it if the draugr ignite it.

The gate will slam shut behind you. Defeat the draugr and head to the second floor via a staircase in the north of the room. You will find Daynas Valen's body with his notes and the Ivory Dragon Claw.

Use the Claw to lower the bridge. You will find your first puzzle there.

❖ Solving The Folgunthur Lever Puzzle

This hallway is a fairly common puzzle. The path forward is blocked by four metal gates that are controlled by four levers. It is complicated by the levers activating different pairs of doors. If you are struggling, pull the levers in the order indicated below.

Door	
3	-
2	1

Another puzzle awaits in the next room.

❖ Solving The Folgunthur Pillar Puzzle

This is a fairly unusual puzzle. You enter a room with two hallways and a mechanism that only allows you to enter one at a time. The first room leads to a trio of engravings. The sequence, starting with the one closest to the door, is below.

1. Snake

2. Fish

3. Bird

Look for a lever by one of the thrones to access the other room. You find a trio of stone pillars. You have to spin the pillars to match the above order.

The only remaining obstacle is a Dragon Claw Door, similar to the other ones you have encountered. The combination, from the outermost ring to the innermost one is:

1. Bird

2. Bird

3. Dragon

❖ Defeating Mikrul Gauldurson

The final opponent within the dungeon is Mikrul Gauldurson, one of the three brothers. He is a swordsman that fights using a health-absorbing sword. More draugr will rise to help him fight, making being overwhelmed a real possibility.

The draugr will perish alongside Mikrul, meaning that they will die as soon as you kill him.

Once you defeat him, make sure to take his unique sword and Gauldur Amulet Fragment. A Word Wall at the end will provide you with a word for Frost Breath.

Searching Geirmund's Hall

A man runs into a cave with a rich autumn forest around him.

Your investigation will bring you to a long-buried cave at the lake to the east of Ivarstead. The entrance is an unassuming cave in the center of it. Defeat some skeevers and look out for a circular hole in the cave to drop into the actual ruin.

You will need to defeat more draugr and Frostbite Spiders throughout the dungeon It is a relatively straightforward encounter that will not be much of a challenge for you.

Solving The Pillar Puzzle At Geirmund's Hall

Your progress will be stopped by a locked door that needs you to rotate four pillars to match a sequence indicated earlier in the dungeon. If you are stuck, the solution is below.

1.	Bird
2.	Fish
3.	Snake
4.	Fish

The next room has several draugr, including a pair located on an upper floor that is difficult to attack from a distance. It would be easiest to run towards the raised area at the end of the room where you can find Geirmund's body. Take the key from its hands to open the locked door opposite. You will be able to attack the remaining draugr.

❖ Defeating Sigidis Gauldurson

You fight one of the brothers, Sigidis, as your final opponent. He is an annoying archer that uses magic to create clones of himself as a form of camouflage. He will only take damage if directly hit.

There are two differences between Sigidis and the clones. He is on the left in the image.

The clones have a blue glow around their silhouettes.

The clones' helmets lack horns. Sigidis' horns point downwards.

Make sure to loot his unique bow and Gauldur Amulet Fragment before you leave.

Jyrik Gauldurson And Saarthal

The final brother, Jyrik, can only be accessed through the College of Winterhold questline. His tomb, Saarthal, is explored early on in the story. It has its own share of puzzles and other interesting mechanics.

We cover how to reach the end and defeat Jyrik in our guide for Under Saarthal.

Reaching Reachwater Rock

The next and final stage is to combine all three fragments back into their completed form. You will have to visit Reachwater Rock, behind a waterfall near Markarth.

You will see a dead adventurer next to the Emerald Dragon Claw shortly after entering. Pick it up and use it to open the first door. The combination, from the outermost ring to the innermost ring is below.

1. Bear
2. Fish
3. Snake

This is immediately followed by a second door that uses the Ivory Dragon Claw. It has the same combination.

1. Bird
2. Bird
3. Dragon

Defeating The Brothers

You will need to confront the three brothers, one after the other, to complete the quest. They will repeat the same mechanics that they first used against you with little variation.

Mikrul will fight you first, raising a small group of draugr to serve as his minions. Sigidis will try to defeat you after Mikrul yields. He will use the same trick, though you can only use the horns as a guide. Finally, Jyrik will make his attempt.

With the three brothers defeated, their father, Gauldur, will appear. He reforges the amulet into its full glory before disappearing. The quest is completed.

The Gauldur Amulet increases your Health, Stamina, and Magicka by 30 points each. It is a versatile artifact that suits well-rounded builds.

DAWNGUARD

MAIN QUESTS

Dawnguard

❖ Starting Dawnguard

You automatically start Dawnguard when you reach level ten. At this point, you might be approached by an orc named Durak. He will ask if you know anything about the vampire threat and if you would like to do anything about it. Regardless of your answer, he will suggest signing up at Fort Dawnguard to put a stop to the threat.

Alternatively, once you reach level ten, guards might randomly comment that the Dawnguard is reforming at the old Fort near Riften.

It is also possible to join by entering Dayspring Canyon before you reach level ten.

Once prompted, the quest Dawnguard will begin. Its quest marker will point to Dayspring Canyon to the southeast of Riften. You will need to take the path wrapping around the city walls to find a road leading to Dayspring Canyon.

Joining The Dawnguard

A pair of recruits listen to a pair of veterans discuss their plans in a stone castle.

When you arrive, you will meet Agmaer, another hopeful recruit. He asks you to walk up with him to calm his nerves. Following the dirt path through the canyon will eventually get you to the entrance to the Fort.

You will see Durak practicing with his Crossbow on the way up. You can ask him what it is. He will give you a Crossbow of your own and a handful of Bolts to use it with.

Crossbows are a unique feature of Dawnguard. They ignore 50 percent of enemy armor and use Bolts instead of Arrows.

When you enter the Fort itself, you will find Isran, the leader of the Dawnguard, speaking to Vigilant Tolan about something going on at Dimhollow Crypt. Turning to speak to you, Isran will ask why you joined before telling you that your first assignment is to investigate Dimhollow Crypt.

Adding that you can take whatever you need to prepare, there are a handful of items in the empty Fort for you to use.

1. A few Bolts and Arrows by the shooting range at the north wing of the Fort.

2. Food items by the kitchen and dining room in the west wing of the Fort.

3. Dawnguard Boots, Dawnguard Shield, and Dawnguard Weapons scattered

across the Roof

The Fort will come to life shortly and will have more for you to explore.

The Dawnguard Weapons deal bonus damage to vampires, making them very useful for the quests to come.

This starts the next quest in the storyline, Awakening.

If you have already decided that you want to side with the Volkihar Clan, don't worry, that chance will come soon.

Awakening

❖ Starting Awakening

This quest starts immediately after the conclusion of Dawnguard. Acting quickly on the information provided by Vigilant Tolan, Isran immediately sends you to investigate Dimhollow Crypt near Morthal. It is in the mountains to the east of the town in an inaccessible location.

Walking directly to the quest marker from Morthal will lead you to a sheer cliff face Instead, follow the road northeast to the Hall of the Vigilant to find the path up.

The cave is close to the destroyed Hall of the Vigilant and has the path toward it littered with corpses. It is a long ascent up the mountain before you reach the entrance.

Entering Dimhollow Crypt

As soon as you enter, you will hear a pair of vampires talking about killing a Vigilant. Vigilant Tolan's body is by their feet. The two vampires are hostile and will attack you as soon as they notice your presence.

Vigilant Tolan is carrying a Potion of Cure Disease that you should pick up. It can cure you of Sanguinare Vampiris, the disease that transforms you into a vampire.

Remember to use it at the end of the dungeon as you will be fighting many vampires that can spread it to you.

The path deeper into the Crypt is blocked by a metal gate. To open it, turn around and look for the Nordic tower on the other side of the entrance. Pull the chain on the second floor to raise the gate and enter.

Dimhollow Crypt is a fairly conventional Skyrim dungeon with the exception of the many vampires filling its halls. There are few surprises as you fight through it.

If you are unfamiliar with fighting vampires, they have a few tricks that you should be aware of.

Invisibility	Vampires, especially higher-level ones, can turn invisible when

	threatened. They often do so before running away before they are defeated.
Vampiric Drain	Vampires will often use a spell to drain your health throughout a fight, healing themselves in the process.
Magic	Vampires are powerful spellcasters in general and will reanimate corpses, summon familiars, and use offensive spells against you.

At the end of Dimhollow Crypt, you will see a powerful vampire fighting a Wounded Frostbite Spider. Beyond this gate is the next and final area, Dimhollow Cavern.

You will hear a vampire, Lokil, taunt a Vigilant before killing him. End the vampire and his allies to continue to the final puzzle.

❖ Solving The Dimhollow Cavern Puzzle

Walk to the central platform and place your hand on the pedestal. After getting your hand speared, a purple glow will come from the furrows in the ground. To solve this puzzle, you need to push the braziers to the end of each glowing line.

After doing this four times, the puzzle will be complete, and the artifact will emerge As it happens, it is a person called Serana, a vampire carrying an Elder Scroll. She asks you to bring her home, ending the quest and starting Bloodline.

Bloodline

❖ Prerequisites To Complete

There are two prerequisite quests that you need to be complete before starting Bloodline. Both of these quests can be found below.

Quest	NPC To Speak With
Dawnguard	Agmaer Located in Dayspring Canyon, east of Riften Alternatively, you just need to be level 10 and this quest will automatically be accepted
Awakening	Isran Located in Fort Dawnguard, east of Riften

❖ Accepting Bloodline

During the prerequisite quests, you will visit Dawnguard Fort and learn of the vampires who attacked the fort.

After venturing to Dimhollow Crypt in search of these vampires, you meet Serana, who will automatically give you the Bloodline quest once Awakening is complete.

❖ Exiting Dimhollow Crypt

Once accepting Bloodline, Serana will ask you to lead her home to Castle Volkihar, located to the north of Northwatch Keep. After speaking with Serana about returning home, Gargoyles will appear around you and Serana, which must be defeated.

With the Gargoyles defeated, head to the exit of the cavern room that you are in and enter into an area with a lever and a gate. Pull this lever and proceed through the gate, while killing the draugr that appear.

Continue on through the gate and enter into an arena-like room, filled with enemies. Kill all of these enemies and leave through the iron door, but be sure to approach the word wall and learn the Drain Vitality shout.

After you leave Dimhollow Crypt, you can make your way to Castle Volkihar.

❖ Traveling To Castle Volkihar

As mentioned earlier, Castle Volkihar is located north of Northwatch Keep. This is an old fortress where the Thalmor keep Thorald Gray-Mane as a prisoner. You can reach this fortress by heading west from Solitude.

On the northern edge of Northwatch Keep, you will find Icewater Jetty with an empty boat. This boat serves as your transportation the first time you visit Castle Volkihar.

Board the boat and head to the castle, where you will meet Lord Harkon.

❖ Speaking With Lord Harkon

Here, you will watch Serana and Lord Harkon reunite, as he questions her about the Elder Scroll she has, as well as who you are.

Eventually, Lord Harkon will present you with the opportunity to become a vampire. If you accept the offer, Harkon will bite you, which causes you to pass out.

If you refuse the offer to become a vampire, you will be banished from Castle Volkihar and won't be able to enter back inside. From here, you will need to return back to Isran and finish the quest.

Later during the Dawnguard questline, you will have the opportunity to become a vampire again during the Chasing Echoes quest. This is helpful if you change your mind after refusing Harkon.

❖ Vampire Lord Training

After Harkon turns you into a vampire, he will give you some training so you can learn the basics of being a vampire lord. This training will occur in the castle's cathedral. Below, you can find the tasks that you will need to perform during this training session.

- Transform into a Vampire Lord
- Land on the ground
- View Perk Tree
- Select a Power

❖ Rewards For Completing Bloodline

perk tree for vampire with all perks active

The only reward you will receive for completing Bloodline is the option to become a vampire. This means that if you refuse the offer, you will not receive any reward.

If you become a vampire, you will also have access to the Vampire Lord Perk tree. This is a series of perks unique to vampires that can be unlocked as you perform vampire activities. For example, as you consume enemies or use the Drain Life spell, you will rank up your vampire ability, allowing you to access new perks.

Once this quest is complete, you will start one of the following quests.

- A New Order
  - this quest will occur if you refuse to become a vampire
- The Bloodstone Chalice
  - this quest will occur if you become a vampire

The Bloodstone Chalice quest will be provided by Garan Merethi, who is located within the castle. If you returned back to Isran and refuse to become a vampire, he will automatically provide you with A New Order.

A New Order — If you sided with the Dawnguard

❖ Prerequisites To Complete

Before beginning A New Order, you will need to complete the following quests.

Quest	NPC To Speak With

Quest	NPC To Speak With
Dawnguard	Agmaer Located in Dayspring Canyon, east of Riften. Alternatively, you just need to be level 10 and this quest will automatically be accepted.
Awakening	Isran Located in Fort Dawnguard, east of Riften.
Bloodline	Serana Located In Dimhollow Crypt.

As mentioned earlier, you cannot be a vampire if you want to do this quest. While completing Bloodline, Lord Harkon will present you with the opportunity to become a vampire. During this conversation, you should refuse his offer and return to Fort Dawnguard.

❖ Accepting A New Order

To accept A New Order, speak with Isran, located at Fort Dawnguard. While doing so, a group of vampires will attack the fort. Your first task is to defend Fort Dawnguard from these vampires.

❖ Recruiting Gunmar and Sorine Jurard

After the vampires have been dealt with, Irsan will ask you to recruit a few people for him. He will request that you find the Nord hunter Gunmar, and Sorine Jurard, a Breton expert on Dwemer technology.

❖ Locating Gunmar

To find Gunmar, you will need to check his location on the map. This will be random as he is not guaranteed to appear at one specific spot.

Once you find Gunmar, he will ask you to help kill a bear that he has been tracking. This will be very straightforward; find and kill the bear, then Gunmar will head back to Fort Dawnguard to help Isran.

❖ Locating Sorine Jurard

You can find Sorine Jurard south of Darkfall Cave. While speaking with Sorine, you can immediately persuade her to go to Fort Dawnguard. If this works, you do not have to do anything else.

If this doesn't work, then you will need to find her lost satchel that was full of Dwemer gyros. You will find this satchel at the riverbank, near the tree rooted in a rock. Alternatively, if you cannot find her lost item, you can also bring her any Dwemer gyro.

Once Sorine is satisfied, she will head to Fort Dawnguard.

❖ Returning To Fort Dawnguard

After both Gunmar and Sorine have been recruited, make your way back to Fort Dawnguard and speak with Isran. Here, he will test both recruits to make sure that they are not vampires.

Both characters will pass the test, proving that they aren't vampires, and Isran will explain the current situation with the Volkihar Clan to them. Once he is done speaking to Gunmar and Sorine, Isran will call you upstairs, which will begin the next quest.

Now that you have successfully completed A New Order, head to Forebears' Hideout for the Prophet quest.

Prophet

❖ Prerequisites To Complete

Although this quest varies depending on your alignment, the prerequisite quests will remain the same. Below, you can check out the quests you will need to complete before accepting Prophet.

Quest	NPC To Speak With
Dawnguard	Agmaer Located in Dayspring Canyon, east of Riften. Alternatively, you just need to be level 10 and this quest will automatically be accepted.
Awakening	Isran Located in Fort Dawnguard, east of Riften.
Bloodline	Serana Located In Dimhollow Crypt.

156

Quest	NPC To Speak With
A New Order If you refuse to become a vampire.	Isran Located in Fort Dawnguard, east of Riften.
The Bloodstone Chalice If you become a vampire.	Garan Merethi Located in Castle Valkihar.

❖ Accepting Prophet Quest

To accept the quest, you will need to meet with the following people.

- If you are a Vampire:
  - Meet Lord Harkon in Castle Vokihar.
- If you are a Dawnguard:
  - Meet Isran at Fort Dawnguard.

❖ Speaking With Isran Or Harkon

Once you spoke with Harkon as a vampire, you will need to listen to his speech about ending the sun's tyranny. After this strange speech, he will tell you to locate a Moth Priest to read the Elder Scrolls.

Rather than speaking with Harkon, you will speak with Isran if you are a member of the Dawnguard. Here, he will talk with you and Serana. Serana will inform the two of you about Harkon's plans, and that he needs to be stopped. To do so, you will need to locate and rescue the Moth Priest as well.

❖ Locating The Moth Priest

To find the priest, you can ask a carriage driver that can be found outside any city. He will have information on the priest's whereabouts, but you will need to either persuade or bribe him to gain it. If successful, the carriage driver will tell you that he took the priest to Dragon Bridge.

You can also speak with an innkeeper in any city, as well as Urag gro-Shub in the College of Winterhold. Both of these options will point you to Dragon Bridge as well.

❖ Arriving At Dragon Bridge

Once you make it to Dragon Bridge, you will need to continue asking about the Moth Priest's locations. While doing so, you learn that the priest isn't there, but he recently crossed the bridge heading south.

Head south until you come to a wrecked carriage with a dead vampire. This vampire is carrying a note mentioning that the Moth Priest is in Forebears' Holdout, which is located to the northeast of your current location.

If you can't find Forebears' Holdout, don't worry; all you need to do is follow the bloodstains on the ground.

❖ Inside Forebears' Holdout

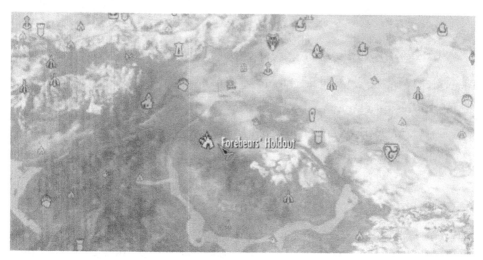

While inside the Holdout, you will encounter several enemies, including members of the Dawnguard or vampires depending on your alignment. Kill any monster, vampire, or Dawnguard that stands in your way and head towards the Moth Priest, named Dexion Evicus. There will be a barrier around the priest that needs to be disabled.

⬦ Disabling The Barrier As A Vampire

The barrier can be deactivated by using a Weystone found on the corpse of Malkus, a dead orc vampire. Use this Weystone and disable the barrier.

⬦ Disabling The Barrier As A Dawnguard

When you enter the cave that Dexion is being held in, you can either kill all the nearby vampires or pickpocket the Weystone from Malkus. Use the Weystone to disable to the barrier around Dexion Evicus.

158

❖ Meeting Dexion Evicus

Dexion is under a spell, so you will first need to fight him.

◦ *Using Vampire Seduction*

Once he submits, calm him with Vampire's Seduction and then bite him. This is an unlockable vampire power, but if you don't have it yet, you will temporarily be able to use this power. After doing this, you will command him to return to Castle Volkihar.

◦ *Rescuing The Moth Priest*

If you are a member of the Dawnguard, you will rescue Dexion after he submits. Speak with him for a bit, until he agrees to read the Elder Scroll. After this, he will head to Fort Dawnguard to meet Isran.

❖ Finishing The Quest As A Vampire

Finally, it's time to make your way back to Castle Volkihar. Inside, you will find the usual group of vampires standing alongside Dexion Evicus. Talk with Harkon for a bit, and then order Dexion to read the Elder Scroll.

After the scroll has been read, the quest will end, and you will receive Chasing Echoes and Seeking Disclosure.

❖ Finishing The Quest As A Dawnguard

Head back to Fort Dawnguard, where you will find Isran and Dexion. The priest will begin to read the Elder Scroll, and the quest will end. Once this is done, you will also receive Chasing Echoes and Seeking Disclosure.

Seeking Disclosure

❖ Prerequisites To Complete

Seeking Disclosure runs concurrently with Chasing Echoes and Beyond Death. If you have yet to retrieve the Dragon Elder Scroll, it will also run concurrently with Elder Knowledge. Below, you can check out every quest in the Dawnguard DLC that needs to be completed before starting Seeking Disclosure.

Quest	NPC To Speak With
Dawnguard	Agmaer Located in Dayspring Canyon, east of Riften. Alternatively, you just need to be level 10 and this quest will automatically be

Quest	NPC To Speak With
	accepted.
Awakening	Isran Located in Fort Dawnguard, east of Riften.
Bloodline	Serana Located in Dimhollow Crypt.
A New Order (If you refuse to become a vampire).	Isran Located in Fort Dawnguard, east of Riften.
The Bloodstone Chalice (If you become a vampire).	Garan Merethi Located in Castle Valkihar.
Prophet	Harkon (if vampire). Located In Forbears' Holdout. Isran (if Dawnguard). Located In Forbears' Holdout.

❖ *Accepting Seeking Disclosure*

To accept the quest, visit one of the NPCs listed below depending on which side of the DLC you are on

- Harkon
    - If you are a Vampire
- Isran
    - If you are in the Dawnguard

Upon speaking with one of these people, you will be given Seeking Disclosure, which requires you to bring the Dragon and Blood Elder Scroll to Dexion Evicus. Although the NPCs are in different locations, the location of the scrolls will remain the same. Next, let's look at where you can find these scrolls.

❖ *Blood Elder Scroll*

The Blood Elder Scroll will be rewarded to you once you complete Chasing Echoes and Beyond Death. These two quests require you to enter into the Soul Cairn, 'slay' Durnehviir, and take the Elder Scroll.

Once you accept Seeking Disclosure, speak with Serana and she will mention the possible location of the scroll. You will not be able to complete this quest until you finish Chasing Echoes and Beyond Death; this is the only way to obtain the Blood Elder Scroll. For more information, be sure to check out our guide on completing Beyond Death.

❖ Dragon Elder Scroll

The Dragon Elder Scroll can be obtained through two different methods, which are listed below.

◇ Elder Knowledge

This quest is part of the main storyline of the game. Elder Knowledge is part of Act II, consisting of locating and recovering the Dragon Elder Scroll.

This is a fairly long quest that is accompanied by the side quest Discerning The Transmundane. To retrieve the Elder Scroll, you will need to complete both of these quests. Fortunately, the quests intersect, and you will be able to complete them at the same time. Overall, if you play through the main storyline of Skyrim, you will be able to obtain the Dragon Elder Scroll.

◇ Scroll Scouting Quest

If you no longer have the scroll from Elder Knowledge, then you will need to complete Scroll Scouting. For this quest, you will need to visit Urag gro-Shub at the College of Winterhold and buy the scroll back for 4,000 gold.

It's important to note that you can only start scroll scouting if you once had the Elder Scroll, but then sold it to Urag gro-Shub. If you have never had the scroll in your possession, then you will need to complete Elder Knowledge first.

❖ Bringing The Scrolls To Dexion

With both Elder Scrolls in your inventory, it's time to visit Dexion. After showing him the scrolls, he explains that he has been blinded, and won't be able to read the scrolls. Instead of reading them for you, Dexion mentions that you should travel to Ancestor Glade and perform the Ritual of the Ancestor Moth. Here, you should be able to read the scrolls and learn the location of Auriel's Bow. Before leaving, Dexion will give you the Sun Elder Scroll as a reward for completing the quest.

This marks the end of Seeking Disclosure, leading directly into Unseen Visions, where you need to enter Ancestor Glade and read the scrolls. Now that you have both scrolls, head to Ancestor Glade and continue on through the Dawnguard storyline.

Chasing Echoes

❖ *Prerequisites To Complete*

Below, you can find every quest that you need to complete before accepting Chasing Echoes.

Quest	NPC To Speak With
Dawnguard	Agmaer Located in Dayspring Canyon, east of Riften. Alternatively, you just need to be level 10 and this quest will automatically be accepted.
Awakening	Isran Located in Fort Dawnguard, east of Riften.
Bloodline	Serana Located In Dimhollow Crypt.
A New Order (If you refuse to become a vampire).	Isran Located in Fort Dawnguard, east of Riften.
The Bloodstone Chalice (If you become a vampire).	Garan Merethi Located in Castle Valkihar.
Prophet	Harkon (if vampire). Located In Castle Volkihar Isran (if Dawnguard). Located In Fort Dawnguard
Seeking Disclosure You will need to accept this quest, but it cannot be finished until you retrieve the Elder Scroll at the end of Beyond Death.	Harkon (if vampire). Located In Castle Volkihar. Isran (if Dawnguard). Located In Fort Dawnguard.

To accept Chasing Echoes, speak with Serana. After completing Prophet, you will either be in Castle Volkihar or Fort Dawnguard. Fortunately, Serana will be nearby regardless of where you are. Approach and speak with her to begin Chasing Echoes.

❖ Exploring Castle Volkihar

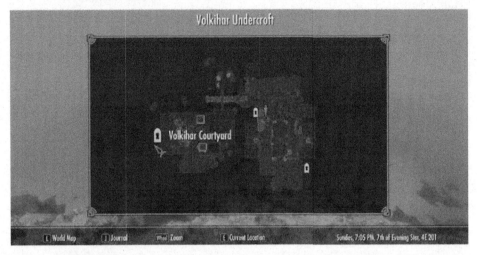

Upon speaking with Serana, she will inform you that her mother, Valerica, has information on the Elder Scroll. To obtain this information, the two of you will need to explore Castle Volkihar searching for any clues.

To start searching, head to a hidden entrance on the back of the castle that is guarded by skeletons. This is the way into the Volkihar Undercroft. Slay these enemies and proceed inside until you come to several other enemies, including a feral Vampire and Death Hounds.

Once these enemies have been dealt with, head up the staircase in the room that leads to a lever.

Pulling this lever will lower a bridge in the lower portion of the room. With the lever pulled, proceed through the passage that opened and take a left on the raised bridge into a large room filled with more Death Hounds.

In this area, you will find another lever that is guarded by a frostbite spider. Slay this spider and pull the lever, which will reveal a door leading to the courtyard. Kill any enemies that stand in the way, and then proceed into the courtyard.

❖ Fixing The Moondial

In the courtyard, you will find a broken moondial. Serana will mention how beautiful this area used to be before her mother left, and you will be tasked with fixing the moondial.

There are three missing pieces that need to be retrieved from the area. Below, you can find the location of each crest that is missing.

Crest	Location
Half Moon	This crest is in a small body of water located near the moondial.
Crescent Moon	On the east of the courtyard, you can find a balcony that contains this crest.
Full Moon	Below the balcony mentioned above, you will find a fenced-off garden that holds this crest.

Gather all three crests and place them around the moondial. Doing so will cause a secret stairway to appear, leading to the Castle Volkihar Ruins.

❖ Inside The Castle Volkihar Ruins

The castle ruins are separated from the rest of the main castle and are filled with enemies such as skeletons and gargoyles. Within the ruins, you will find rooms similar to the main castle, however, it will be a bit more run down. Follow the main path through the ruins, taking the stairway into a dining room. Here, kill all the enemies that stand in your way and then continue to the south balcony until you come to a room with traps and a gargoyle.

Continue forward and head through the ruins, avoiding any traps and slaying enemies that appear. Serana will be at your side, providing occasional directions and slaying enemies.

Eventually, you will come to a room with a fireplace. This fireplace is a secret entrance (Serana will tell you this as well), and you need to find how to enter. This can be done by either turning a candlestick near the fireplace or pulling a chain found behind a gargoyle. Both options will open the fireplace, revealing a hidden study.

❖ Valerica's Study

In Valerica's study, you will be instructed to find her journal, which can be found in a bookcase to the right of the entrance. Serana will read this, and then tell you to gather three ingredients.

Below, you can find the location of all three ingredients around the study. Each ingredient will be in a large silver bow.

Ingredient	Location
Finely Ground Bone Meal	This is located on a table by the entrance, near a mammoth skull.
Soul Gem Shards	This is located on the first wardrobe after heading up the staircase.
Purified Void Salts	This is located on an ingredient shelf, located on a balcony above the entrance.

After gathering all the ingredients, place them in front of the portal that Serana is standing near. Once placed, Serana will add her blood and the Soul Cairn portal will start to open. Once it's fully glowing purple with a pathway exposed, you can enter.

❖ Entering the Soul Cairn Portal

If you are a vampire, you can proceed into the Soul Cairn portal and finish Chasing Echoes.

If you are not a vampire yet, you will have two options. To enter the Soul Cairn, you must be dead, so you can either accept Serana's offer to transform you into a vampire, or have her trap part of your soul in a soul gem.

When Serana turns you into a vampire, you will gain all the powers of a Vampire Lord, and will no longer be accepted by the Dawnguard. If your soul is trapped in the soul gem, you will remain human but have less health, stamina, and magicka, as well as reduced regeneration rates while in the Soul Cairn. These stats will return to normal once you leave the Soul Cairn after completing Beyond Death.

You do not have to be a Vampire Lord to enter the Soul Cairn without using a soul gem. If you are affected by the normal vampirism disease, you can still enter. This is helpful if you want to remain aligned with the Dawnguard. After exiting the Soul Cairn, locate an innkeeper or member of the Dawnguard, who will give you instructions to curing your vampirism.

Choose one of these options and then proceed into the Soul Cairn. Once you enter, Beyond Death will begin, tasking you with finding the Elder Scroll and confronting Durnhviir.

Beyond Death

❖ *Prerequisites To Complete*

Beyond Death is near the end of the Dawnguard questline. Below, you can find all quests that you need to complete first.

Quest	NPC To Speak With
Dawnguard	Agmaer Located in Dayspring Canyon, east of Riften. Alternatively, you just need to be level 10 and this quest will automatically be accepted.
Awakening	Isran Located in Fort Dawnguard, east of Riften.
Bloodline	Serana Located In Dimhollow Crypt.
A New Order (If you refuse to become a vampire).	Isran Located in Fort Dawnguard, east of Riften.
The Bloodstone Chalice (If you become a vampire).	Garan Merethi Located in Castle Valkihar.
Prophet	Harkon (if vampire). Located In Forbears' Holdout. Isran (if Dawnguard). Located In Forbears' Holdout.
Seeking Disclosure You will need to accept this quest, but it cannot be finished until you retrieve the Elder Scroll at the end	Harkon (if vampire). Located In Castle Volkihar. Isran (if Dawnguard).

Quest	NPC To Speak With
of Beyond Death.	Located In Fort Dawnguard.
Chasing Echoes	Serana. Located In Castle Volkihar.

⋄ *Accepting Beyond Death*

After completing Chasing Echoes, this quest will immediately be accepted and added to your journal. The Chasing Echoes ends with you entering the Soul Cairn, so you will already be in the location for Beyond Death.

❖ Locating Valerica

Your first task in Beyond Death is to locate Serana's mother, Valerica, within the Soul Cairn. The path to her is pretty straightforward; continue into the cairn while following the waypoint marker.

While heading towards Valerica, you will encounter several lost souls, with some of them giving you side quests for the Dawnguard.

❖ Speaking With Valerica

Once you reach Valerica, you will notice a barrier blocking her from reaching you and Serana. Speak with Valerica here, and she will tell you that the only way to remove the barrier is to defeat the three Boneyard Keepers. These are enemies that can be found on three towers within the Soul Cairn.

❖ Traveling To The Boneyard Keepers

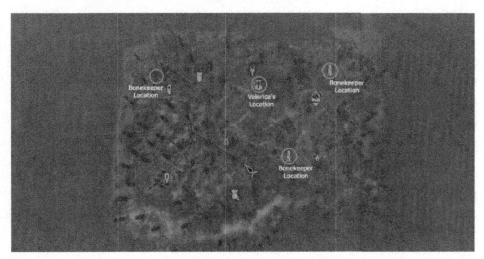

The tower locations are pretty far apart from each other. However, if you complete the Soul Cairn Horse Quest, you will be able to summon Arvak, a horse mount. By summoning Arvak, you can quickly travel around the cairn.

In the image above, you can find the location of each Keeper, as well as Valerica's location. As you approach each Keeper, you will come across weaker enemies. Take care of them, and then proceed to the Keeper. Killing the Keepers will not be difficult if you have made it through the rest of the quests in the Dawnguard questline with little to no trouble.

If you are struggling to defeat the Keepers, try to position yourself in such a way that the use of Unrelenting Force will push the Keeper off of the tower. Doing so will cause the Keeper to fall and die.

❖ Returning To Valerica

After all three Boneyard Keepers have been defeated, head back to Valerica. Once you return, the surrounding barrier will vanish, and she will lead you into the Boneyard. Here, Valerica will give you your last task: slay Durnehviir.

❖ Defeating Durnehviir

Durnehviir is an undead dragon that can be found within the Soul Cairn. Slaying this dragon is relatively easy compared to previous boss' battles.

The best method to take down Durnehviir is by using a bow and arrow (or a crossbow with bolts). Using this long-ranged weapon, shoot Durnehviir as he flies through the sky. Once the dragon lands, Valerica will begin to attack him. This will distract Durnehviir, allowing you to continue attacking him with a long-ranged weapon.

Continue attacking Durnehviir, as well as the minions that he summons, until he is defeated. Since he is the undead dragon, Durnehviir doesn't actually die after this fight is over.

❖ Retrieving The Elder Scroll

After you defeat Durnehviir, follow Valerica to the location of the Elder Scroll, and then exit the Boneyard.

Serana and Durnehviir will be waiting outside of Boneyard, waiting to speak with you. Here, Durnehviir will teach you a shout that can summon him, as well as request that you summon him in Tamriel, so he can see where he once lived.

❖ Finishing Beyond Death

Lastly, head back through the portal and back to Tamriel; doing so will end the quest. The only reward that you will receive from this quest is the Dragon Shout that allows you to summon Durnevhiir.

With both the Blood and Dragon Elder Scroll in hand, it's time to head to Dexion and complete Seeking Disclosure.

Unseen Visions

❖ *Prerequisites To Complete*

Below, you can find all the prerequisite quests that you will need to complete, along with the quest giver and their location.

Quest	NPC To Speak With
Dawnguard	Agmaer Located in Dayspring Canyon, east of Riften Alternatively, you just need to be level 10 and this quest will automatically be accepted
Awakening	Isran Located in Fort Dawnguard, east of Riften
Bloodline	Serana Located In Dimhollow Crypt
A New Order if you refuse to become a vampire	Isran Located in Fort Dawnguard, east of Riften
The Bloodstone Chalice if you become a vampire	Garan Merethi Located in Castle Valkihar
Prophet	Harkon (if vampire) Located In Castle Volkihar Isran (if Dawnguard) Located In Fort Dawnguard
Seeking Disclosure	Harkon (if vampire)

Quest	NPC To Speak With
	Located In Castle Volkihar Isran (if Dawnguard) Located In Fort Dawnguard
Chasing Echoes	Serana Located In Castle Volkihar
Beyond Death	This quest will automatically be added to your journal

◦ *Accepting Unseen Visions*

After completing Seeking Disclosure, Dexion Evicus will automatically give you this quest. Your first task will be to head to Ancestor Glade.

❖ *Traveling To Ancestor Glade*

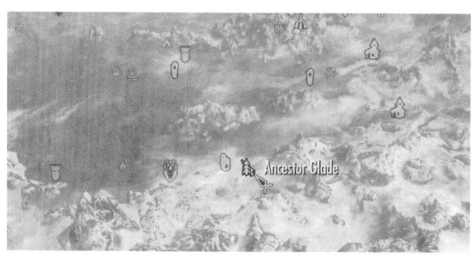

Ancestor Glade can be found near Falkreath. Follow your waypoint marker from this location until you arrive at the glade. If you are struggling to find Ancestor Glade, you can find a step-by-step walkthrough below.

1. Fast travel to Falkreath, and then proceed out the eastern road.

2. On this road, you will come to an intersection. Take the northeastern path,

which leads to Pinewatch.

3. Pinewatch is a bandit outpost, with a small mountain path nearby marked with stones. Head up this mountain path until you come to a small bandit camp.

4. From this camp, head right up a path until the ground turns to snow.

5. At the end of this path, you will be met with a rock wall, with a pile of rocks to the left.

6. Go past this rock pile and head right until you come across another rock pile with an attached flag.

7. Follow the path with flagged rocks until you reach the entrance of Ancestor Glade.

8. The entrance of the glade can be found in a cave, to the left of the path that you are on.

Once you make it here, enter the cave and head into the cavern. When your first enter, there will be no enemies that you need to deal with, so you can take your time and look around. The caves leading to Ancestor Glade are fairly straightforward; just follow the designated path until you arrive in a lush cavern filled with trees.

❖ Find A Moth Priest's Knife

Within Ancestor Glade, you will need to find a Moth Priest's draw knife. This can be found at the center of the glade, laying on a circular stone. A draw knife is a saw-shaped tool used in woodworking to shave pieces of wood and in this case; tree bark.

Take this draw knife and collect bark from a nearby tree. Doing so will allow you to collect swarms of moths. Collecting the moth swarms is simple, just explore around the glade and walk through the moth swarms. After seven swarms have been collected, head to the beam of light in the glade.

❖ Reading The Elder Scrolls

In the beam of light, you will read all three Elder Scrolls that you have on hand. Your quest will instruct you to read the Blood Elder Scroll, but upon doing so, all three will be read. After reading the scrolls, a vision of Auriel's Bow will be presented to you, and then your vision will white out. Although this is the location of the bow, you will not need to remember it. Once this quest is complete, a waypoint will lead you there.

❖ Speaking With Serana

Wake up and head to Serana and speak to her about what you saw. Once the conversation is over, the quest will complete and you will automatically begin

Touching The Sky.

Prior to beginning Touching The Sky, you will need to deal with the enemies that spawn around you after the conversation with Serana. Quickly take care of these enemies, and then make your way to Darkfall Cave to begin Touching The Sky.

Touching The Sky

❖ *Prerequisites To Complete*

As part of the Dawnguard questline, you will need to progress through the following quests before you can begin Touching The Sky.

Quest	NPC To Speak With
Dawnguard	Agmaer Located in Dayspring Canyon, east of Riften Alternatively, you just need to be level 10 and this quest will automatically be accepted
Awakening	Isran Located in Fort Dawnguard, east of Riften
Bloodline	Serana Located In Dimhollow Crypt
A New Order if you refuse to become a vampire	Isran Located in Fort Dawnguard, east of Riften
The Bloodstone Chalice if you become a vampire	Garan Merethi Located in Castle Valkihar
Prophet	Harkon (if vampire) Located In Forbears' Holdout Isran (if Dawnguard) Located In Forbears' Holdout

Quest	NPC To Speak With
Seeking Disclosure	Harkon (if vampire) Located In Castle Volkihar Isran (if Dawnguard) Located In Fort Dawnguard
Chasing Echoes	Serana Located In Castle Volkihar
Beyond Death	This quest will automatically appear in your journal
Unseen Visions	Dexion Located in Ancestor Glade

If you follow the story of this DLC, either as a member of the Dawnguard or as a vampire, you should not have any trouble accepting Touching The Sky.

⋄   *Accepting Touching The Sky*

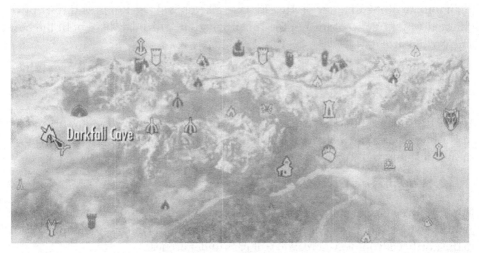

To accept this quest, you will need to meet Serana, who can be found in Darkfall Cave. Once you speak to her, this quest will begin. Darkfall Cave is located south of Castle Volkihar. Additionally, you can get here by heading north from Markarth.

❖ Inside Darkfall Cave

Your first task in this quest is to locate Auriel's Bow within Darkfall Cave. As the name suggests, the cave will be very dark. To aid in your search, you can light a torch, use your vampire sight (if you are a vampire), or use light-casting spells.

Within the cave, you will find frostbite spiders and trolls. These enemies are easy to deal with, so quickly defeat any that stand in your way.

As you make your way through the cave, you will come across a small wooden bridge. Here, you will need to jump off the bridge and land in the water below. Follow the current and proceed forward until you come to a campsite area and a fork in the road.

If you chose to go the path behind the dead Breton woman, you will encounter a tripwire trap that releases a boulder. Be cautious of this boulder and continue onward, killing the two trolls that stand in the way.

The other path will lead you to an area filled with spiderwebs. On this path, you will encounter a giant frostbite spider, as well as two trolls at the end of the passage

Both of these passages will lead to the same place, where you will see the Knight-Paladin named Gelebor praying at an altar devoted to Auriel.

❖ Speaking With Gelebor

Gelebor will help you find Auriel's Bow, in exchange for your help killing his brother Vyrthur, who has been corrupted by Falmer.

This will not be a simple task though; to kill Vyrthur, you must first follow the path of an initiate of the Chantry of Auri-El. Overall, you will need to pass through five different wayshrines that lead to the Inner Sanctum, which is where both the bow and Vyrthur are.

Additionally, you will need to fill the Initiate's Ewer with water from all five shrines.

❖ Wayshrine Of Illumination

This is the first area that Gelebor teleports you to. Here, you will follow a winding road while slaying any monsters that stand in your way. Getting to the wayshrine is simple; just follow the path until you come to the shrine.

❖ Wayshrine Of Sight

Now, progress farther until you exit the passage into the Forgotten Vale. This is a giant snowy area, but you can quickly reach the second shrine by going up the path to the north. On your way here, you will pass under arches.

❖ Wayshrine Of Learning

This shrine can be found to the west of the arches mentioned above. From the arches, head west up the stairs until you come to a narrow path. Kill any frostbite spiders that stand in your way, and then drop down into the river below.

You can find this shrine as you head downstream. After collecting water from the Wayshrine of Learning, head up the steep path leading to the frozen lake.

❖ Wayshrine Of Resolution

The fourth shrine can be found on the northwest bank of the frozen lake. On your way here, you can also find part of the shout for Drain Vitality on a rock that sticks out into the frozen lake.

❖ Wayshrine Of Radiance

It's now time to visit the final shrine. From the Wayshrine of Resolution, head north to the Glacial Crevice. This is a long path that will take you through the living areas of the Falmer. Continue through this narrow passage until you come to a clearing high on a cliff with the shrine. Grab the water with your Ewer, and proceed to the Inner Sanctum where Vyrthur and the bow are.

❖ Inside The Inner Sanctum

Once you make it to the Inner Sanctum, enter inside and place the Ewer on an altar to open the doors. After the doors open, be sure to grab the Ewer.

Proceed deeper into the Inner Sanctum until you make it to the chapel. As you enter, you will face three waves of enemies. At the end of the third wave, Vyrthur will cause the ceiling to collapse on you. After passing out, Serana will wake you up and instruct you to go to the balcony. Make your way here and confront Vyrthur.

⬩ Killing Vyrthur

Vyrthur will use ice magic, Vampiric Drain, and Lightning Bolt. Additionally, he will summon a frost Atronach.

Overall, Vyrthur shouldn't be too hard to handle. Dodge his attacks while dealing damage in return, and he should be dead in no time.

❖ Completing Touching The Sky

Now that Vyrthur is dead, a wayshrine will appear and Gelebor will walk out of it. Speak with Gelebor and claim Auriel's Bow for yourself. Once you do this, the quest will end, and you can begin Kindred Judgement.

⬩ Rewards For Completion

As a reward for completing Touching The Flames, you will be rewarded with the following items.

- Auriel's Bow

- 20 Elven Arrows
- 12 Sunhallowed Arrows

Auriel's Bow is fairly strong, dealing 20 points of sun damage to enemies. If shot at an undead target, this will become 60 points of sun damage. Gelebor mentions that the bow was carried by Auri-El himself, and that "its craftsmanship has no equal anywhere within Tamriel".

Now that you have Auriel's bow, it's time to go confront Lord Harkon in Kindred Judgement.

Kindred Judgement

❖ Prerequisites To Complete

Below, you can find the quests you need to complete before starting Kindred Judgement. This is the last quest in the questline, meaning that this list consists of every quest included in the questline.

Quest	NPC To Speak With
Dawnguard	Agmaer Located in Dayspring Canyon, east of Riften Alternatively, you just need to be level 10 and this quest will automatically be accepted
Awakening	Isran Located in Fort Dawnguard, east of Riften
Bloodline	Serana Located In Dimhollow Crypt
A New Order if you refuse to become a vampire	Isran Located in Fort Dawnguard, east of Riften
The Bloodstone Chalice if you become a vampire	Garan Merethi Located in Castle Valkihar

Quest	NPC To Speak With
Prophet	Harkon (if vampire) Located In Forbears' Holdout Isran (if Dawnguard) Located In Forbears' Holdout
Seeking Disclosure	Harkon (if vampire) Located In Castle Volkihar Isran (if Dawnguard) Located In Fort Dawnguard
Chasing Echoes	Serana Located In Castle Volkihar
Beyond Death	This quest will automatically be added to your journal
Unseen Visions	Dexion Located in Ancestor Glade
Touching The Sky	Serana Located in Darkfall Cave, south of Castle Volkihar

⋄ *Accepting Kindred Judgement*

Kindred Judgement will be automatically accepted once you pick up Auriel's Bow in the quest Touching The Sky. With the quest active, your first task is to speak with Serana.

After speaking with her, you will need to do one of the following quests.

- If you are a Vampire:
    - Travel to Castle Volkihar
- If you are Dawnguard:
    - Travel to Fort Dawnguard and meet with Isran

At this point in the story, you will be able to fast travel to both Castle Volkihar and

Fort Dawnguard, allowing you to quickly get to each location.

❖ Meeting With Isran

This section will only apply to players that have aligned with the Dawnguard. If you are a vampire, you can skip through this section.

Once you speak with Serana, head to Fort Dawnguard and speak with Isran. Here, he will give a speech, mentioning that it's time to attack the vampires and kill Harkon.

From here, make your way to Castle Volkihar. Since you are not a vampire, you will need to fight your way inside the castle, slaying gargoyles that block the bridge leading to the front doors. In addition to the gargoyles, you will need to defeat vampires that will emerge from the castle.

If defeating these vampires proves to be difficult, you can shoot a Sunhallowed arrow towards the sun with Auriel's Bow. This will cause the effects of the sun to deal damage to the vampires, but this only works if it is daytime.

After these enemies have been defeated, head into Castle Volkihar and defeat the vampires that appear. These vampires are strong, but you will have the option to watch while the rest of the Dawnguard fights the vampires. If you choose to ignore these vampires and continue, they will not follow you.

Make your way to the Cathedral of the castle, where Harkon is waiting.

❖ Entering Castle Volkihar As A Vampire

If you are a vampire, you can proceed directly into the castle and head straight to the Cathedral. Vampires will be walking around the castle, but they will not be hostile. The Dawnguard will not be present at this time, and you do not have to engage in any fighting.

❖ Speak With Harkon

Now that you have made it to the Cathedral of Volkihar Castle, it's time to speak with Harkon. As you enter this area, Harkon will be in his Vampire Lord form and will begin to talk with Serana.

Harkon will eventually ask you to hand Auriel's Bow to him. Although you have this option, it's not recommended to hand it over. The bow will help during the fight with him, so if you get rid of it, you will be at a huge disadvantage.

After speaking with Harkon, it's time to engage in the final battle of the Dawnguard questline.

❖ Fighting Vampire Lord Harkon

The fight against Harkon is the same, regardless of whether you are aligned with the Dawnguard or vampires. Overall, the fight will be hard, so be sure to bring any healing item necessary.

At the start of the battle, Harkon will summon skeletons and gargoyles. Kill these smaller enemies right away so they will not get in the way as you fight Harkon. Additionally, Serana will help you fight these enemies.

Harkon will use the Bats Night Power to quickly move around the room while casting Vampiric Drain towards you.

As he takes more damage, he will transform into a swarm of bats and return to the altar in the Cathedral. Here, he will create a shield to protect him while he regenerates his health. This shield can be destroyed by firing any type of arrow from Auriel's Bow.

This mechanic happens several times; whenever he appears above the altar, shoot his shield and continue to fire arrows at him.

Once his health is low enough, Harkon will travel around the room in a mist form, materializing from time to time. Follow the mist and attack whenever Harkon appears.

Eventually, Harkon's health deplete to zero. When this happens, he will transform into a swarm of bats, appear by the altar, and melt into red ash.

❖ Completing Kindred Judgement

After Harkon is dead, you will either be approached by Isran or Garan Merethi. Isran will offer for you to become a member of the Dawnguard if you are not a vampire.

If you are a vampire, Garan Merethi will speak to you and Castle Volkihar will become your home. This means that everything in the castle is yours, and you can come and go as you please.

❖ Rewards For Completing Kindred Judgement

skyrim_harkon's_sword_inside_inventory

As a reward for completing this quest, you will receive the following items.

- Harkon's Sword
  - If you are a vampire, wielding this sword will absorb 15 points of Magicka, Health, and Stamina from enemies
- Vampire Royal Armor
  - Wearing this causes Magicka to generate 125% faster
- Potion of Blood
- Random Jewelry
- Random Potions

All of these items can be looted directly from Harkon's pile of ashes.

If you are a vampire, Serana will become a permanent follower. Additionally, she will have the ability to create Bloodcursed Elven Arrows with her blood. These arrows have the ability to turn the sun red and prevent negative effects from the sun on vampires until the following day.

That's all there is to know about the Kindred Judgement quest! Now that you have completed the main questline for the Dawnguard DLC, you can start the sidequests provided by the vampires and Dawnguard.

SIDE QUESTS

Lost To The Ages

❖ Discovering The Quest

There are three ways to initiate this quest. Whichever route you take, this quest isn't assigned its own name right from the get-go. You'll find it under the Miscellaneous tasks.

1. Find and read a copy of The Aetherium Wars (Miscellaneous task: "Investigate the ruins of Arkngthamz")

2. Discover an Aetherium Shard before reading Aetherium Wars. (Miscellaneous task: "Identify the Crystal Shard.") If this was the way you discovered the quest you'll need to do one of the other two methods in order to know where to go.

3. Find and enter Arkngthamz with no understanding of what you're getting yourself into. (Miscellaneous task: "Investigate the ruins of Arkgthamz")

The first thing to do is reach Arkgthamz, which is in The Reach, south of Markarth. If you found it already, you'll be lucky enough to just teleport. If it's a new location for you, fast-travel to your nearest identified location and start running - or trotting! It's a good idea to bring a horse, since its gravity-defying powers will help you to traverse the rocky landscape much quicker.

❖ Inside Arkgthamz

Upon entering Arkgthamz, you'll start to hear a ghostly voice. Continue on until you meet Katria, the ghost of a Nord who died in these halls. If you insist that you need to push on, she'll decide that she cannot convince you not to and offers her help. Agree to take it! As you cross the chasm in front of you, check the little island for Katria's body and loot it to get her journal.

Push through the pipes to reach a flooded room and climb up the pipe along the wall to get to the next section. In the room where Katria comments on the Dwarven machine that you won't find in Skyrim clans, attack the rotating mechanism to open the next door.

Progress through the tunnels, killing Falmer as you go. When you pass through the Chaurus encampment, you'll want to exit by the right-hand side. The next area you reach will include a few trees and bushes, with a toppled log on one side. At the end of that log is Katria's bow, Zephyr, which you can pick up and use - it fires 30% faster than other bows.

❖ Tonal Lock

The final room asks you to solve a puzzle. There are rotating mechanisms like the ones you've seen before, built into the wall. Katria will stop you to explain that this is a deadly test called a "Tonal Lock" and failing it is what caused the earthquake that killed her. You must hit the mechanisms in a precise order. Katria's journal and a scrap of paper found on a dead adventurer provide the clues you'll need.

This is the order:

1. Bottom Left

2. Bottom Right

3. Top Left

4. Top Right

5. Bottom Centre

Adding Zephyr to this dungeon is a clever way to give you a bow for this part of the quest. If you're out of arrows, just go back and loot some of the Falmer. Alternatively, projectile spells also work.

The door will unlock and you will get access to the final room, full of Dwarven treasures to loot - and an Aetherium Shard. Katria explains that this is the key she spent her life looking for, made out of pure Aetherium, but that it's been split into four pieces. Your next task will be to find the other three! The quick exit to this dungeon is hidden in the nearby river, just past the skeleton with a spear coming out of its chest.

❖ Tracking Down The Shards

Note that the quest markers on your world map for tracking down the Shards are notoriously buggy. They may appear as soon as you leave Arkgthamz, they may appear when you've gotten close enough to one of them, or they may not appear at all. If in doubt, refer here for help.

◊ Deep Folk Crossing

The Deep Folk Crossing Shard is probably the easiest to obtain. Deep Folk Crossing is a Dwemmer Bridge crossing a river in the far north of The Reach.

Cross the bridge, walking north, and then follow the slope up to find a little Dwarven plinth with a Dwarven Sword and Helmet on one side and an Aetherium Shard on the other.

◊ Mzulft

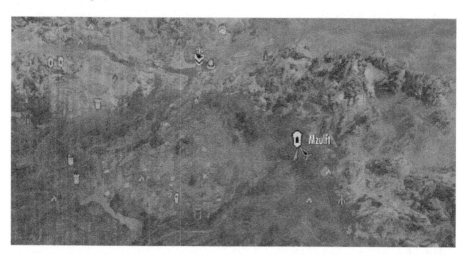

The next Shard is at Mzulft, a Dwemer ruin in Eastmarch, southeast of Windhelm. You may recognize the name; that's because it's a critical part of the College of Winterhold storyline. No need to worry if you haven't done the College of Winterhold yet. The main section won't be available to you, but it doesn't matter because you're entering a side room. This might be the Shard you've already found, if you were really adventurous during the College quest.

Its door is labeled as a "Dwemer Storehouse." Inside, Katria will be waiting for you. She'll ask for help unlocking a door - you can either unlock the "Expert" lock in front of her, or unlock two "Apprentice" locks to circumvent it. Inside, past those doors, along with Dwarven goodies, is the Aetherium Shard.

◆ *Raldbthar*

The third Shard is in a Dwemer ruin called Raldbthar. If you found one of these Shards in advance, it might have been here, since the player progresses through it during the Dark Brotherhood quest "Mourning Never Comes."

These caves are riddled with bandits, Falmer, and Dwarven machines. Once you arrive and push your way through the first through enemies, you'll enter another area that's called "Raldbthar Deep Market."

There are several puzzles to solve as you venture forth. The first is a room with four buttons - you need to hit the button second from the right. Next, there will be a room with a raised bridge. You need to lower it, but there's no power. To restore power, check all the gears for blockages (i.e. a Human Spine) - and don't forget the one underwater!

You should be able to press forward easily now. Your next challenge is the giant Dwarven Centurion. Once it's finally dead you'll progress into the final set of rooms and eventually see the Aetherium Shard sitting alone on a pedestal. Assuming the Quest Marker is working, let that be your guide.

❖ The Forge

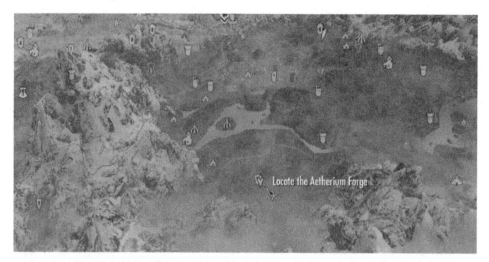

When you finally have all four Shards, venture to the Forge. There should be a quest marker on your map, but you can also rely on Katria's notes. It's marked as a location on the map as the "Ruins of Bthlaft". Be sure to bring a bow or some other projectile spell with you, since you need to deal with a few more rotating devices. Push forward to find the Aetherium Forge.

You'll need to deal with the Dwarven countermeasures; there will be a few waves of Dwarven Spiders and Dwarven Spheres before a boss fight with a Dwarven Centurion called "The Forgemaster."

❖ Your Reward

Once you've taken it down, you'll finally have the opportunity to use this gorgeous forge. Grab the "optional" crafting items off to the side of the room and then interact with the forge. There are three possible items you can craft:

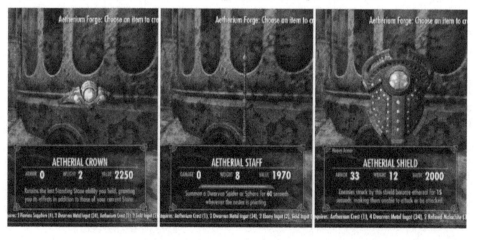

184

- an Aetherial Crown, which allows you to take on two different Standing Stone abilities

- an Aetherial Shield, which turns enemies ethereal

- an Aetherial Staff, which spawns a Dwarven Spider of Sphere ally

Katria will say a brief goodbye and thank you for helping her with her unfinished business in the mortal realm. Investigate the room, if you want (spells and shouts like Become Ethereal will help), before departing!

How To Marry Serana

❖ Who Is Serana in The Elder Scrolls: Skyrim?

Serana is a vampire featured in the Skyrim expansion Dawnguard. You'll first meet Serana during the "Awakening" quest in the Dimhollow Crypt while investigating the Hall of the Vigilant attack. After setting her free, Serana will ask you to help her get home to Castle Volkihar in order to find out what happened to her mother Valerica.

Serana is the daughter of Valerica and Lord Harkon, the main antagonist in the Dawnbreaker DLC. Serana is one of the few known pureblood vampires, and as such, is titled one of the "Daughters of Coldharbour."

Lord Harkon discovers a prophecy called the Tyranny of the Sun, which has led him to believe that by collecting a series of artifacts, including two Elder Scrolls, he can purge sunlight from the world, giving vampires the ability to walk freely day and night. There's just one wrinkle: he needs to sacrifice his daughter in order to do so.

❖ How To Get Married In Skyrim

In order to get married in Skyrim, you must first complete the quest "The Bonds of Matrimony." To receive the quest, speak with Maramal in the Temple of Mara. Maramal will instruct you to procure an Amulet of Mara.

Maramel will sell you the amulet for 200 gold, but there are other ways to obtain one. A priestess named Dinya Balu at the Temple of Mara will give you a quest called "The Book of Love." At the end of the quest, you will have an Amulet of Mara you can use to get married. You can also find an amulet in random loot and shops.

Once you have the amulet, identify the person you would like to marry, and speak to them while wearing the amulet to propose. (Simple times, eh?) Once you've done so, head back to the Temple of Mara to arrange the wedding. Leave the temple and wait until you receive the objective "Attend your wedding." Then, return to the temple and the wedding will begin.

❖ Why Can't You Marry Serana in Skyrim?

Marriage has numerous benefits in Skyrim, including a daily home-cooked meal, a store that provides passive income, and occasional gifts from your spouse. Not to mention the role-play advantages of getting married, as doing so will advance your personal story. It's a meaningless metric to some, but an invaluable one for many others.

Though your options for marriage are many, including male and female characters and a dozen choices of race, there is one limitation in Skyrim that will prevent you from marrying Serana: you cannot marry vampires.

While it is possible to turn your spouse into a vampire in the Dawnguard expansion once you yourself have become a vampire, it isn't possible to marry someone who is already a vampire, like Serana. Fortunately, this is where fan mods come in to pick up the slack and give players what they want.

❖ How To Add The Mod That Lets You Marry Serana in Skyrim

The mod, simply titled "Marriable Serana," was created by Codrmonk33 and is available on NexusMods. Here are the steps to take if you want to marry Serana on PC:

- You need to purchase and install the Dawnguard expansion in order for the mod to work, so do that first.

- Download "Marriable Serana" by clicking "manual" next to download on NexusMods.

- Unzip the file and add it to your Skyrim folder, usually found at C / Program Files (x86) / Steam / steamapps / common / Skyrim and move the mod folder into Skyrim's data folder.

If you're playing the game on a PS4 or Xbox One version (or their next-gen counterparts), you can still mod your file accordingly. On PS4/PS5, search for "Marry Me Serana" within the game's mods menu. On Xbox One/Series, you ought to be able to use the "Marriable Serana" mod as-is (and you can look it up either within the mods menu in-game or on Bethesda's website).

Marrying Serana works just like marrying any other NPC in Skyrim, so after proposing, go to the Temple of Mara to set up the wedding - and that's it. You're now married to the spookiest lady in Skyrim. Kill the father, marry the daughter, what more could you want out of life?

Find Arvak's Skull

❖ Who Is Arvak?

Arvak is a horse that you can encounter within the Soul Cairn in the Dawnguard DLC. Unlike other steeds in Skyrim, you can summon Arvak using a spell, which is exclusive to this mount. Arvak is a skeleton, but he can still function in the same way as any other horse.

❖ How To Learn The "Summon Arvak" Spell In Skyrim

First, you need to enter the Soul Cairn, which you can do by following the "Chasing Echos" quest during the Dawnguard DLC's main storyline — you'll need to complete this regardless of whether you side with the Dawnguard or Volkihar Vampires. Become a vampire before entering the Soul Cairn or allow Serena to partially trap your soul inside a Soul Gem.

When you enter the Soul Cairn and start walking down the main path, you'll encounter a soul shouting about Arvak. He'll plead with you to find Arvak, but you'll first need to retrieve Arvak's Skull to summon the horse.

Along this path, you may encounter enemies of a few different varieties, so make sure to bring a weapon and a powerful suit of armor. Luckily, most of the enemies are few and far between, but they're most common at landmarks and structures throughout the Soul Cairn.

You can discover Arvak's Skull east of the location where you find Valerica; at the location in the picture above. You may need to defeat a few Mistmen, Wrathmen, and Bonemen before picking up Arvak's Skull, but most of the enemies are quite weak, so you can easily defeat them with any ranged or melee weapon.

Once you pick up Arvak's Skull, a soul will appear and speak to you about the spectral steed. He'll teach you the "Summon Arvak" spell, and you can now conjure Arvak whenever you want. You can find out more about summoning Arvak below.

❖ How To Summon Arvak In Skyrim

Once you learn the "Summon Arvak" spell, you can go into the Conjuration menu and equip the spell. The base Magicka cost of "Summon Arvak" is 136, so that's the only requirement you need for casting it.

Once you summon Arvak, you can ride your new horse in any outside location in Skyrim, Solstheim, or the exterior locations of Blackreach and the Soul Cairn. Like regular horses, you can't summon or ride Arvak inside dungeons, buildings, or certain cities.

DRAGONBORN

MAIN QUESTS

Dragonborn

❖ *The Cultist Attack*

To start Dragonborn, you need to have unlocked the main story quest, The Horn Of Jurgen Windcaller. A pair of Cultists will begin searching Skyrim for you, likely finding you at a settlement in the daytime. After asking if you are Dragonborn, they will call you a fake and try to kill you.

The Cultists are dangerous enemies, especially if you are early in the game. They have access to offensive magic and can summon a Flame Atronach.

❖ *Starting Dragonborn*

Killing them will start the quest Dragonborn, asking you to read Cultists' Orders, a note you find on one of their bodies. It will tell you that they had come from Solstheim by way of Windhelm.

You will need to visit the Windhelm Docks to find Gjalund Salt-Sage, the sailor that had brought the Cultists to Skyrim. He is clearly distressed and confused about the Cultists and initially refuses to go back to Solstheim. You have to convince him to bring you there or pay double his usual rate.

You will be able to travel directly without having to convince him and by paying less gold if you go to Solstheim before starting Dragonborn.

❖ *Arriving At Raven Rock*

With passage secured, you make it to the Dark Elf settlement of Raven Rock. The town is run down and desolate. Immediately after you disembark, you are questioned by Adril Arano, an official representing House Redoran about the purpose of your visit. Asking him about Miraak will confuse him, though he will mention the Earth Stone, updating the quest, and placing a marker on the Stone.

Speaking to the Orc Mogrul will let you skip the visit to the Earth Stone.

On your arrival, you will see locals constructing a temple while mumbling to themselves and unable to give you any useful information. You will be approached by a man named Neloth who is observing the workers. Asking him about Miraak will have him point the way to The Temple of Miraak.

While tempting, it would be best to explore Raven Rock later in the expansion as the Earth Stone is currently disturbing the town's activities. Returning later would be advised.

The temple itself is to the northeast of Raven Rock and getting there will take you on a journey through Solstheim. Once you arrive, you will find another temple being constructed by mind-controlled villagers. Frea, a warrior calling out to her friends will be there. Talking to her will start the next quest.

The Temple Of Miraak

❖ *Starting The Temple Of Miraak*

This quest starts as soon as you speak to Frea at the end of Dragonborn. Unlike most of the people you've met, she's from the Skaal, an indigenous tribe unique to Solstheim. Her mind had been preserved by a necklace designed to protect her from the mind-controlling magic.

The two of you will be attacked by a pair of Cultists emerging from the area underneath the Temple. While challenging enemies, they are easily defeated. Head down the path they came from to find the entrance to the Temple of Miraak.

❖ *The Temple Of Miraak*

Immediately after entering, Frea will stop to investigate the Temple. Between the corpses laying about and hanging in gibbets, she has plenty to talk about. You can take this opportunity to loot the area, or you could forge ahead. Either way, she will come with you.

Traveling into the Temple, the two of you reach a room with a large pit at the center. Take the stairs down to encounter more enemies in the form of cultists and Draugr.

Frea notes that there is some treasure beyond a broken staircase.

To get it, climb up what remains of the staircase, turn right, and hop up to the supporting pillar. Use Whirlwind Sprint to reach the chest.

There will be a hallway lined with swinging axes next. While Whirlwind Sprint still works, the hallway is much longer than the ones you've previously encountered. You will need to pace yourself and shelter at the gaps between sets of axe blades. It may take several uses of Whirlwind Sprint to reach the end.

The Shout Become Ethereal or simply evading the axes with skill both work as well.

Your path will be blocked in the next room by a raised drawbridge. To lower it, head to the doorway opposite the entrance, entering an alcove facing the bridge. There is a handle that will lower the bridge there.

In the next and final room in this area, three drawbridges will fall in dramatic fashion as you enter. A powerful Cultist and two draugr will descend to look for you Defeat all of them and go to the door at the back to reach Temple Sanctum.

❖ The Temple Of Miraak Sanctum

There will be more draugr and Cultists waiting for you deeper into the Sanctum. Be mindful of traps waiting to be triggered close to enemies. Clever movement will turn the traps against your enemies.

Try to stay close to Frea. She will often comment on the surroundings, giving hints about nearby treasure and dropping interesting nuggets of information about Solstheim's history.

You will reach a Word Wall for Dragon Aspect. As soon as you claim it, the coffins lining the room will open and a mixture of normal and challenging draugr will attack as a group. One of them is called the Gatekeeper. It carries the Temple of Miraak Key that will open the remaining doors in this dungeon.

Dragon Aspect can only be used once a day. It may be tempting to use it immediately, but its benefits may be more useful elsewhere.

The next area is a dining room and an adjoining kitchen. Frea will stop to comment on the two rooms and remark that it is a dead end. To proceed, walk through the kitchen, head to the northwest passageway across from the entrance, and take the first turn right. The handle to reveal a hidden passage is in the alcove at the far end.

Another blocked path awaits you. Go to the room on the right and pull another handle. This opens a passage down. Follow it and pull another handle at the dead end to reveal another secret doorway. This leads to a large room dominated by a pair of stairs with many draugr.

Watch out for traps set up along the stairs and rocks rolling down. While seemingly harmless, the rocks can quickly kill you if you are unlucky.

Fortunately, they only roll down one side at a time. Go to the other side to avoid them.

The final enemy is an immensely powerful draugr. Kill it and loot the chest by its throne. Pull the chain behind its throne to open a final passageway.

❖ Finishing The Temple Of Miraak

You will enter an area with bizarre, alien-looking architecture. At the very end is a Black Book called Waking Dreams. Interact with it to enter another dimension and meet Miraak. He mocks you, announces his evil scheme, and orders his minions to send you back.

Speak to Frea to share what you have learned. This ends the quest and starts the next one, The Fate of the Skaal.

The Fate Of The Skaal

❖ Starting The Fate Of The Skaal

The quest starts automatically after finishing The Temple of Miraak. Frea will lead you out of the Temple and to Skaal Village to the east. The journey itself isn't very long, and there will be few enemies to contend with as you travel there.

Skaal Village is not usually so empty. Continue the quest to see the village come back to life.

The village itself is nearly empty. The remaining inhabitants are huddled in the center and channeling magic into a barrier to protect them from Miraak. Frea asks you to tell her father, Storn what you have learned. Concerned, he directs you to Saering's Watch, a ruin far from the village to learn a Word of Power that he believes will be vital to stopping Miraak.

❖ Saering's Watch

You will have to travel northwest to reach the Word Wall. As you arrive, you will encounter a dragon attacking draugr. Entering the fray will start a three-sided battle between the dragon, the draugr, and yourself.

While you could enter through the main entrance, you could climb the outer wall instead, jumping down to claim the Word and running out while the enemies are distracted.

You need to find the Word Wall high up in the ruins. It would be advisable to kill the dragon first while the draugr can keep it distracted before focusing on the undead.

Miraak will steal the dragon's soul once you kill it, mocking you in the process and preventing you from adding another soul to your collection.

He will continue to do this until you complete the expansion.

❖ Freeing The Villagers

With the Word learned, you need to go to the Wind Stone and use the Bend Will Shout on the stone. It will trigger a reaction, making it explode while freeing the villagers from Miraak's control.

There will be no time to rest on your laurels. A Lurker will emerge from the temple. This is one of Dragonborn's new enemies. It has three main ways of attacking.

Stomp	It will stomp its foot, creating a writhing swarm of tentacles around it that deal damage over time.
Poison Spit	It will spit poison at a distant enemy, dealing damage over time.
Melee Attacks	It will swipe at you with its claws, damaging your health.

This quest ends when you return to Skaal Village and speak to Storn. This will start the quests Cleansing The Stones and The Path Of Knowledge.

Cleansing The Stones

❖ *Starting Cleansing The Stones*

You begin after reporting your success to Storn. Having freed his people from Miraak's control, he identifies four more Stones for you to cleanse, found throughout Solstheim.

Beast Stone	Found to the east and between the Temple of Miraak and Thrisk Mead Hall
Earth Stone	It is to the west of Raven Rock. You are likely able to fast travel directly to it.
Water Stone	It is to the northwest of Raven Rock.
Sun Stone	In the south-east near the Dark Elf settlement of Tel Mithryn

Each stone cleansed will free more locals, allowing them to return to their usual schedules. Many areas that were once affected by Miraak's control such as Raven Rock will return to normal with many merchants and quest-givers available.

❖ The Beast Stone

This Stone is most easily accessed by fast traveling to the Temple of Miraak and heading southeast. You will find it being attended to by the diminutive Rieklings, a native tribe that has taken over Thrisk Mead Hall nearby.

You can accept the quest The Chief of Thrisk Hall by visiting it after cleansing the stone to learn more about the Rieklings and gain their assistance.

Using Bend Will on the stone will free them from Miraak's control. The Rieklings will cower while a lurker erupts from the Stone.

❖ The Earth Stone

The quickest way to reach it would be to fast travel directly to it. You would have visited it if you did not speak to Mogrul in Dragonborn. The Stone has enthralled both Raven Rock citizens and a handful of Reavers.

Using Bend Will will cause one lurker to appear from the half-constructed temple and another to approach from the shore. The Redoran Guardsmen will help you fight both of them.

The Reavers may begin fighting the Guardsmen immediately after being freed. It may be confusing but helping the Guardsmen would prevent any citizens from being killed.

❖ The Water Stone

Being situated north of the Earth Stone, traveling directly to it would be most convenient. This method may involve jumping up a mountain. A small group of sailors is working on the Water Stone.

Traveling along the coast will bring you to Bloodskal Barrow, which you can get in the quest The Final Descent. It is one of the strongest swords in the game.

The cleansing follows the same pattern as the previous ones, though a dragon may attack while you cleanse the stone. It will be hostile to both you and the lurker.

❖ The Sun Stone

This stone is located in the southeast of Solstheim and is near Tel Mithryn, where you will need to go for the quest The Path Of Knowledge, which is automatically accepted with Cleansing The Stones. Doing this prior to continuing The Path Of Knowledge would be advised.

When cleansing this Stone, you will need to fight three lurkers as well as the Reavers who were enthralled by the Stone.

The Path Of Knowledge

❖ Starting The Path Of Knowledge

This quest starts after freeing the Skaal from the Wind Stone and returning to Storn Crag-Strider at Skaal Village. He tells you to seek out another Black Book, starting The Path of Knowledge, and to free the rest of Solstheim, starting Cleansing The Stones.

The Sun Stone, one of the objectives in Cleansing The Stones, is on the way to Tel Mithryn and can be completed concurrently with this quest.

He suggests visiting the mushroom tower at Tel Mithryn, on the southeast corner of Solstheim to search for the researcher Neloth, who spoke to you in Dragonborn.

At Tel Mithryn, go to the largest mushroom and use the magical lift to find Neloth. After some discussion, he tells you that the Dwarven ruin of Nchardak has a Black Book that he has been trying to reach. Reasoning that two heads are better than one, he sets off to Nchardak. You can follow him or meet him there when you're ready.

His dialogue will change if you have completed the quest Discerning The Transmundane.

❖ Entering Nchardak

The ruin is not far from Tel Mithryn, visible as you exit the area. Your entry is blocked by a group of Reavers that have made the ruin their temporary home. You will need to kill all of them before you can enter.

With the ruins half-submerged, the Reavers are able to attack from a distance as you run from tower to tower. You will need to pace yourself and be mindful of ranged attacks as you fight.

Neloth is no slouch and will join you in fighting the Reavers and the other enemies.

With the ruins cleared of foes, Neloth will use a Control Cube to open the ruins, allowing you to enter the Nchardak Reading Room.

The Black Book is right at the entrance, locked securely beneath a mechanism that Neloth has been unable to open. He explains that you will need to produce enough steam to activate the mechanism for the book to be released.

The researcher brings you to the Nchardak Grand Chamber where he explains that he needs to collect more Control Cubes to both drain the Grand Chamber and activate the boilers to make steam. He provides you with his Cube and points the way to the Nchardak Workshop where you can find three more.

❖ Nchardak Workshop

Your second Control Cube is at the entrance. You will need to pick it up to solve the puzzles to come. This area has two Control Consoles set up at the east end of the workshop. You will need to use a cube on one of them to drain the workshop enough to access the third Control Cube.

Jump down to the bottom of the workshop to use another Control Console, raising a set of stairs to reach the northern door. Your path will be blocked by a bridge. Use the Command Cube to raise that bridge. This will allow you to reach the third Control Cube.

Dwarven automata will attack in waves as you use the mechanisms. Be alert for them.

The fourth Control Cube is beyond the southern door and blocked by a collapsed staircase. As Neloth explains, you need to raise the Workshop's water level such that you'll be able to swim to the Cube.

Remove the Cubes from the Control Consoles to flood the Workshop. Swim through the door and grab the Control Cube. This will flood the workshop. With the Cubes in hand, return to the Great Chamber.

❖ Nchardak Aqueduct

With Control Cubes, you are able to lower the water level in the Great Chamber enough to reach the next area, Nchardak Aqueduct. You will need to reach the Control Console to the right of the entrance.

The path is obstructed by a set of raised bridges. To lower them, walk to the three buttons above and behind the entrance and press the right button followed by the left button.

Automata will attack as soon as you lower the bridges.

Use a Cube at the Control Console to lower the water level. Neloth will remain behind to raise the water level once you have taken the final Cube.

Walk through the marked room and take the Final Command Cube from its pedestal. As the water level rises, swim out of the aqueduct.

❖ Getting The Black Book

Returning to the Great Chamber, you will have to use a single Cube to drain enough water to expose the Boilers. The remaining four Cubes can be used to power the Boilers, making it possible to take the Book.

You will find the button to release the Black Book to be lit and active. Press the button to unlock the mechanism. Neloth will suggest reading the Book. Doing so will drag you inside, completing this quest and starting the next one The Gardener of Men.

The Gardener Of Men

❖ Starting The Gardener Of Men

You are quite literally dragged into this quest when you read Black Book: Epistolary Acumen at the conclusion of The Path of Knowledge. Greeted by Hermaeus Mora himself, you are told that the knowledge you seek lies within Apocrypha. You will have to find it yourself.

You are able to leave Apocrypha by reading the Black Book. Use it when you are lost or overwhelmed.

You begin this quest in Chapter One, alone on a platform.

❖ Chapter One

This is a relatively simple introduction to the confusing reality within Apocrypha. Here, you will need to interact with the Scrye in front of you. It will create a moving bridge. Walk through the bridge to find a distant book on a pedestal. It is beyond your reach, forcing you to wait for the bridge to align with a second platform with a Scrye.

There are two new hazards introduced in this chapter.

The dark ichor between platforms is highly dangerous and will quickly kill you.

Standing close to the edge of platforms will have tentacles reach out and whip you.

Interact with the Scrye for a second bridge that will take you to Chapter Two.

❖ Chapter Two

This chapter begins with you in a narrow tunnel. Walk to the end to activate another Scrye. As the tunnel unfurls, you will have to fight a lurker and two seekers While you would have fought lurkers in The Fate of the Skaal and Cleansing The Stones, the seekers may be an unfamiliar enemy.

Seekers are crafty foes that will frequently turn invisible and conjure doppelgängers. They are able to drain your Health, Stamina, and Magicka as well.

Activate a new Scrye to open the gate to Chapter Three.

❖ Chapter Three

This chapter is as linear as Apocrypha gets. You will need to fight your way through seekers and lurkers while navigating a tunnel similar to the one from Chapter One. You will find the final book, Chapter Four, and the end.

❖ Chapter Four

A man approaches a book held aloft in a strange contraption while towers of books stretch into the sky behind him.

You will reach the knowledge you seek here. The chapter consists solely of a large platform that has a book at the end. Interacting with it will start a conversation with Hermaeus Mora. Though unpleasant, the Daedric Prince provides you with the second word for Bend Will. Before he leaves, he tells you to have the Skaal give their secrets to him.

With your business concluded, you have the opportunity to choose between three abilities.

Dragonborn Frost	The Frost Breath shout will immobilize enemies.
Dragonborn Flame	A Fire Wyrm will emerge from the corpses of those killed by Flame Breath
Dragonborn Force	Unrelenting Force deals more damage and may disintegrate enemies.

To return to Tamriel, interact with the book. You will be in Nchardak with Neloth. As you exit, you will be attacked by the dragon Krosulhah. Defeat it to continue.

❖ Completing The Gardener Of Men

Return to Skaal Village to complete the quest. With no other recourse, you will need to ask Storn to give Hermaeus Mora the secrets he seeks. The Shaman agrees, willing to put Miraak's defeat above the secrets long hidden from the Daedric Prince.

Against Frea's protests, Storn reads the Black Book. Hermaeus Mora appears and kills Storn, carving the final word of Bend Will on his body. With this, the quest ends, and the next one, At the Summit of Apocrypha starts.

At The Summit Of Apocrypha

❖ Starting At The Summit Of Apocrypha

Continuing right after the conclusion to The Gardener Of Men, you will need to use a Dragon Soul to unlock the final word in Bend Will. With this shout, you will be able to rival Miraak's power and stand a chance against him.

This knowledge had come at a price. Frea's father and Skaal shaman, Storn Crag-Strider had been killed by Hermaeus Mora in exchange for the final Word. At his grieving daughter's urging, read Waking Dreams to re-enter Apocrypha and stop Miraak.

Take Epistolary Acumen from Storn's body before you leave. The book will disappear otherwise.

Once again, reading Waking Dreams will drag you into Apocrypha, starting your ascent to the Summit.

❖ Reaching The Summit Of Apocrypha

As always, you begin at Chapter 1. This time, all you need to do is walk toward the quest marker to reach Chapter 2. There, you are introduced to the main obstacle of the dungeon, the need to collect books to solve a puzzle at the end.

⋄ Getting Boneless Limbs

Walking up the stairs, you will need to defeat two seekers. The way forward will be blocked by a set of retracted stairs. To extend them, turn around and look for the book Boneless Limbs on a pedestal overlooking the stairs. Pick the book up to extend the stairs. This will allow you to reach Chapter 3.

You will be fighting many seekers throughout this quest. Boosting your magic resistance will help you survive them.

⋄ Getting Delving Pincers And Prying Orbs

Apocrypha becomes truly mind-bending from here. This area has a confusing layout. You will need to go through the corridor you start at, defeat a pair of seekers and get the book Delving Pincers.

Turn around and go through a previously closed door and interact with a Scrye at the end. You will enter an open area with some enemies. Look for the next book, Prying Orbs, to the southwest. You will be able to reach Chapter 4 once you have it.

⋄ Getting Gnashing Blades

Chapter 4's architecture changes as you navigate it. Start by walking toward the hallway below the entrance. It will contract and summon seekers once you set foot inside. Leave the way you came to discover a completely different room. The book Gnashing Blades can be found overlooking the exit.

You can find Fonts of Magicka and Fonts of Stamina throughout Apocrypha. Activating them will restore their respective resource over time.

To reach Chapter 5, head through the south hallway and head east. You will find a Scrye at the end. Retrace your steps to find a large chamber with a pool of ichor in the middle. A lurker will emerge from it to fight you. Activate a nearby Scrye to progress.

In Chapter 5, you will need to place each of the four books on the correct pedestal. The order is indicated by the patterns on each one. Match the book to unlock the book to Chapter 6. If you are struggling, the solution is below

- Boneless Limbs is in the northeast.

- Delving Pincers is in the southeast.

- Gnashing Blades is in the southwest.

- Prying Orbs is in the northwest.

Chapter 6 is the final stage of this quest.

❖ Reaching Miraak

There are two seekers guarding a Word Wall directly ahead of you. The dragon Sahrotaar will swoop down and attack you after defeating them. Use all three words of the Bend Will Shout to sway him to your side. Offering to bring you to Miraak, you hop on his back.

You will be able to tame other dragons with Bend Will. This flight with Sahrotaar is a tutorial for the mechanics for future uses.

You will have to command him to kill several enemies on the way before commanding him to land at Miraak.

❖ Defeating Miraak

Miraak approaches as you land, saying that he will be able to use your soul to escape Apocrypha. Combat starts shortly after. As the First Dragonborn, Miraak has access to Shouts which he makes ample use of, alongside magic and melee attacks. This combination makes him dangerous at all distances.

He is able to deplete your magicka with Shock magic and your stamina with his unique tentacle-like sword.

When you manage to deplete most of his health, he will use Become Ethereal and Whirlwind Sprint to escape and consume the soul of a nearby dragon. Each time he does this, he will be completely restored. He will do this twice before Hermaeus Mora chooses to kill Miraak for trying to escape, ending the battle.

You earn several rewards for defeating Miraak.

1. For completing Waking Dreams, you are able to trade a Dragon Soul to reallocate your perks without needing to Legendary the Skill Tree.

2. You can loot Miraak's body for his unique weapons and armor.

3. You will absorb Dragon Souls from Miraak, including any that he had stolen from you since completing The Temple Of Miraak.

Read the book to return to Solstheim. This ends Dragonborn's main storyline, though there are many more things to discover across Solstheim.

SIDE QUESTS

March Of The Dead

❖ Starting March Of The Dead

The quest starts when you leave Raven Rock through the Bulwark. It is the large, stone wall built to protect the town from ash storms. You will see Captain Veleth of the Redoran Guard, dressed in distinctive Bonemold Guard Armor, fighting off some Ash Spawn. He is near the Old Attius Farm near the bodies of several fallen guardsmen.

You will likely pass through early in the Dragonborn story during the quest Dragonborn.

Once you help him defeat his attackers, he will ask you to search their bodies for clues. Look for a note titled "Declaration of War". Giving him the note will surprise him as the General that signed off on the declaration died 200 years ago. He asks you to go to Fort Frostmoth to investigate further.

It is also possible to start the quest by entering Fort Frostmouth without speaking to Captain Veleth and killing General Falx Carius independently. He will have a note on his body that can be turned in to Captain Veleth to complete the quest.

❖ Searching Fort Frostmoth

The Fort in question is located on Solstheim's southern coast and is directly east of the Old Attius Farm. The years have not been kind to it and the structure is badly damaged. To enter, walk across the coastline and look out for a set of stairs leading up to the entrance.

Your enemies are primarily comprised of Ash Spawn. These volcanic creatures are notable for their resistance to fire and ability to use Flame spells. Avoid using Flame spells against them and try to use Potions of Resist Fire to stay protected.

You will need to reach a door on the western wall to enter Fort Frostmouth. It is safest to clear out the courtyard instead of rushing into the Fort itself. The defenders will attack you in numbers as soon as you enter.

❖ Defeating General Falx Carius

The General is waiting for you deep within the Fort. Head right at the entrance and take the stairs down. Be alert for Ash Spawn emerging from the ash to fight you. Eventually, you will find a room protected by two Ash Spawn and a locked door that requires a key.

This area has several opportunities for getting more loot. The webbed-up doorway to the west leads to a room with albino spiders and explosive Flame-Cloaked Spiders. While dangerous, clearing it out will allow you to access Heart Stone deposits inside.

A master-locked door to the northwest can be opened by a skilled lockpick or with the Fort Frostmouth Key. Searching for it will provide you with more loot and a significant amount of food.

Go through the doorway to the north of the locked door and head west at the junction. There, you will find a knapsack containing the Fort Frostmouth Key and a journal explaining how the General is alive.

With the key in hand, you can go through the locked door and confront the General in combat. He is a difficult enemy to fight and sports a unique hammer that will deal significant damage to you.

The General is helped by several Ash Spawn. It would be best to kite them across the chamber and use the pillars to shelter from Firebolts. Focus on the Ash Spawn as the General lacks ranged attacks and can be kept at arm's length with swift movement.

Once you defeat him, return to Raven Rock and speak to Captain Veleth and inform him of your success. With the General gone, the Ash Spawn no longer threaten Raven Rock. He will be pleased with the news and give you gold as a reward, scaled to your level.

Make sure to loot General Carius' body for the Champion's Cudgel, a unique two-handed warhammer that is equivalent to a glass warhammer with a special enchantment for Chaos damage.

This enchantment has a 50 percent chance to deal Shock, Frost, Fire damage, or a combination of the three. Thanks to the inherent randomness of the weapon, it is useful in most situations by being able to avoid enemy elemental resistances.

The Final Descent

❖ Starting The Final Descent

You start this quest by entering Raven Rock Mine, on the north of the settlement. The mines have been shut down long ago and should be empty. Instead, you find an old Imperial, Crescius Caerellius, arguing with his Dark Elf wife, Aphia Velothi.

You need to speak to Caerellius despite his foul mood. He will tell you that he is searching for his great-grandfather, Gratian Caerellius' Journal, believed to be deep within a dangerous section of the mine. Aphia is trying to dissuade him from investigating. As a compromise, Caerellius gives you a letter and key from his great-grandfather and asks you to search in his stead.

You will be given the letter regardless of how skeptical you are. You will not have to appease him to progress the quest.

Go behind Caerellius to take a path deeper into the mine to continue.

❖ Getting Through Raven Rock Mine

Descend through the empty mine while making sure to travel in the direction of the quest marker. You will eventually find a locked door that requires the key Caerellius gave you. The challenges begin as you go through them.

You will have to fight through several rooms of draugr as you get deeper into the mine. It becomes apparent that miners had dug into a Nordic burial ground and disturbed the dead.

Keep your eyes peeled for a Stalhrim deposit after a circular room with four draugr. This is a rare location that allows you to gather the resource.

Eventually, you will reach a flooded room with a bridge connecting the two sides. To progress, search the left side of the room for a handle. Pulling it will raise a grate blocking the exit.

At this stage, you will fall into a large room where you will find Gratian Caerellius' remains and Journal.

❖ Escaping Bloodskaal Barrow

You will find the unique Bloodskaal Blade by Gratian Caerellius' remains. You will need to use its unique power attacks to escape Bloodskaal Barrow.

Look at the stone door that dominates the cavern. It has a pair of red fissures on its frame. You will need to perform a power attack with the Bloodskaal Blade to produce an arc of red energy. Striking the fissure with an arc of energy will destroy the fissure and start to unlock the door.

You will also have to be mindful of the orientation of each fissure. They will either be vertical or horizontal. They need to be hit with the corresponding power attack to be destroyed. Use standing power attacks to produce a vertical arc of energy for vertical fissures. Sideways power attacks will produce a horizontal arc of energy for horizontal fissures.

The Bloodskaal Blade's enchantment is useful beyond this puzzle. It is also helpful in combat as the arcs function as ranged attacks. They deal damage and are even effective at staggering enemies.

The weapon is improved with silver ingots if you have the Arcane Blacksmith perk. It cannot be disenchanted.

Once you successfully escape that chamber, you will need to get through a hallway filled with swinging axes. While not a new hazard, it is a long corridor that can turn very deadly if you are not paying attention.

❖ Defeating Zahkriisos

The final part of the Barrow is in an empty room with a chest and a Word Wall. You will be attacked by the Shock-themed Dragon Priest, Zahkriisos as soon as you get close enough.

While tempting, do not enter the water. It is too deep to stand in, and you will be left vulnerable when forced to swim inside. It is safest to fight from the edges or in front of the Word Wall.

It can summon Seekers to assist it and uses Lightning Storm, a Master-level Destruction spell that will evaporate your health bar if you lack Shock resistance.

The Bloodskaal Blade can be particularly useful in this fight as it will stagger Zahkriisos, even through walls.

Make sure to loot Zahkriisos for its mask and to claim a word for the Dragon Aspect shout from the wall.

The mask, Zahkriisos, gives you 50 percent more Shock resistance and increases the damage of Shock spells by 25 percent.

❖ Completing The Final Decent

You will emerge into the Reaver-infested part of Bloodskaal Barrow as you leave the dungeon. They are not particularly dangerous and do not need to be eliminated for you to complete the quest.

You only have to bring the journal back to Crescius Caerellius in Raven Rock. He will give you leveled gold for retrieving the journal.

Your efforts will also reopen the Raven Rock Mine. It is a source of rare and valuable ebony ore used to create Skyrim's most powerful armor and weapons.

Served Cold

❖ Starting Served Cold

Captain Veleth will ask you to talk to Adril Arano after completing the quests March of the Dead and Final Descent. In both of these quests, you would have saved Raven Rock from an onslaught of Ash Spawn and reopened their ebony mine returning them their source of income.

Arano was the first person to greet you when you arrived at Raven Rock to start Dragonborn.

Arano is compelled to ask you for help as he is trying to stop his friend and colleague, Lleri Moravyn, from assassination. Thanks to his status and personal friendship, the assassins are watching him closely and are quick to clear their tracks whenever he tries to investigate them.

❖ Identifying The Assassins

He tells you to start by visiting the Retching Netch Cornerclub and speaking to the barkeep, Geldis Sadri. Sadri gladly tells you that the assassins are likely to visit the

Ulen Ancestral Tomb to pay their respects. Previous attempts had ended without any visitors. He hopes that your reputation as a newcomer will let you identify the people involved.

You can find the Ancestral Tomb by the Bulwark. It is unlocked and easily accessed Simply head inside and use the Wait feature to pass a few hours. Tilisu Severin will enter the tomb.

You will need to be close enough to read her name for the quest to consider her identified. While you could do so stealthily, talking to her is equally effective.

Arano will give you the Severin Manor Key once told of her role and ask you to find evidence within.

❖ *Finding The Evidence*

The key gets you into Severin Manor without much difficulty though Tilisu and Mirri Severin will investigate the noise and attack you. You can kill them without any repercussions.

They may not immediately attack if you are skilled in Sneak.

Make sure to search Mirri's body for the Severin Family Safe Key. The safe is found in the bedroom, directly across from the entrance, and down the stairs. It is on the left of the room, by a set of shelves. You can help yourself to anything inside, though you need to take a note titled "The Ulen Matter"

Bring this note back to Adril. He wastes no time, asking you to raid Ashfallow Citadel, where the assassins are waiting.

❖ *Stopping The Assassins At Ashfallow Citadel*

Adril will send two Redoran Guards to the Citadel and asks you to join them. You will be attacked by Morag Tong assassins as soon as you arrive at the citadel. These enemies are fast and dangerous melee fighters. They are so deadly that the guards you are meant to accompany are already dead.

Enter the Citadel to defeat the assassins and foil the plot. The way to the ringleader is blocked by three sets of metal bars that need to be lowered for you to progress. More Morag Tong assassins will attempt to stop you along the way.

The First Set	These metal bars will be lowered by using the pull chain in a room on the left side of the hallway, with an Alchemy Lab inside.
The Second Set	You need to pull two chains. The first is at the end of a narrow corridor on the right of the hall. A second chain is on the left of the bars.

The Third Set	You need to walk through several swinging spike walls by avoiding the pressure plates. You will find two chains on the right side of the room. One deactivates the traps and the other lowers the bars.

With all the bars lowered, you can reach the final room where Vendil Severin is waiting for you with a pair of assassins.

You should try to kill the assassins first, as they will attack you with arrows. Leaving them alone for too long will slowly drain your health while Vendil devastates you.

The final battle is challenging, but you will have completed your objective by killing Vendil.

❖ Completing Served Cold

With the day saved, you return to Arano, who brings you to the Councilor. Thanks to your hard work in restoring Raven Rock's fortunes, you are made a citizen of the settlement.

More importantly, you are awarded a pile of gold and Severin Manor. You have, after all, killed the previous owners.

The Chief Of Thirsk Hall

❖ Starting The Chief Of Thirsk Hall

A group of unusually civil Rieklings will stop you as you pass the hall for the first time and haltingly order you to speak to the Chief inside the hall. Humoring them will get you face-to-face with their leader. Unlike most of his kind, he can converse with you.

You are likely to pass by this area when visiting the Beast Stone in the Dragonborn story quest, Cleansing The Stones.

After a short conversation, he demands you help the Rieklings get their house in order by getting them a Bristleback.

❖ How To Make Bilgemuck Return To Thirsk Hall

His first request is to coax their Bristleback, Bilgemuck, back into its pen. You will find it wandering alone, north of the nearby lake. Approaching Bilgemuck will automatically open a menu for you to offer meat to it. Any kind of meat indicated on the menu will suffice.

Be sure not to run too quickly as Bilgemuck may struggle to follow you if it cannot keep up.

Walk back to Thirsk Hall with the animal in tow. The objective will be automatically completed when it gets close enough to its pen. There is no need to get it directly inside; Bilgemuck will do so automatically.

❖ Where To Get Scathecraw

His second demand will be for ten Scathecraw, a red plant that grows abundantly across Solstheim. You would be able to gather more than enough if you wander about the vicinity of the hall.

You will find a large quantity of the plant growing around Raven Rock's town square, by the blacksmith. A quick trip there will get you all that you need.

The Chief will be pleased with the delivery and make his final request; killing the nearby Nords. The recently-evicted owners had made some attempts to reclaim Thirsk. Their deaths would ensure that his tribe can live in safety. You can agree immediately or ask for some time to think. If you prefer to avoid killing the Nords, make sure to ask for time before searching for them.

Leaving the conversation before giving the Chief an answer may result in the Scathecraw being taken without the objective being complete. This will force you to gather it again.

❖ Killing The Nords

The Chief will give a short speech before leading his tribe down to Bujold's Retreat. You will have to join him in killing all seven Nords. Each one is a competent fighter that will not go down quietly.

The diminutive Rieklings are individually weak and will not be of much assistance. You will be doing most of the fighting yourself.

Once the deed is done, the Chief will feel threatened by your prowess and demand to fight you to the death to preserve his status. You will need to slay him to complete the quest.

Once this is done, you will be able to recruit Rieklings as followers.

❖ Retaking Thirsk Hall With The Nords

You can side with the exiled Nords at any time by speaking to Bujold the Unworthy at her camp, Bujold's Retreat by the coast. She will explain that the Nords of Thirsk Hall had grown soft with a cushy lifestyle and had been driven out by the Rieklings. If you volunteer to help her, she quickly rouses her clan to action.

The Rieklings are no different from the usual variety that you fight. They are easy to defeat.

You will join them in purging Thirsk Hall of its infestation. The Rieklings will try to cling to their new holdings and put up stiff resistance to defend themselves.

With their home reclaimed, Bujold will seek the blessing of Thirsk Hall's founder to continue her leadership. As the person instrumental to their success, she asks you to travel to Hrothmund's Barrow with her as a witness.

❖ The Options For Retaking Thirsk

There is no combat necessary at Hrothmund's Barrow. There, Bujold will be rejected by Hrothmund. She asks you to keep it a secret. You will need to choose how to proceed from here. Refusing to keep the secret will have Bujold fight you while agreeing to help will keep her alive.

Thirsk Hall's members will offer side quests regardless of your choice.

Narrative reasons aside, the only thing at stake is Kuvar's Master-level Heavy Armor Training. Any choice that has him learning that Bujold was rejected will prevent you from accessing it.

A New Source Of Stalhrim

❖ Starting A New Source Of Stalhrim

You can learn about the missing blacksmith when walking through Skaal Village. The chief, Fanari Strong-Voice, is arguing with Deor Woodcutter about searching for the village's missing blacksmith.

You can only begin this quest after completing The Fate of the Skaal, part of the Dragonborn storyline.

Deor wants to search for Baldor Iron-Shaper, the blacksmith, as his sudden disappearance is highly unusual. Fanari does not share his concerns and discourages him from personally investigating.

Talk to Deor after the conversation. He explains that Baldor's knowledge of stalhrim smithing would make him a valuable target for people interested in the secret. He asks you to search for Baldor, suggesting a search southwest.

❖ Finding Baldor Iron-Shaper

The quest marker will direct you to the Abandoned Lodge found north of Raven Rock. You will find a collection of Thalmor operatives waiting outside the lodge. They will attack you if you get too close.

They are a good source of high elf blood for the quest Discerning the Transmundane.

Loot one of their bodies for a key and a note. The key will unlock the lodge, where you find Baldor wounded in the basement. He does not need any further help and tells you that the Thalmor are trying to learn how to use stalhrim and have stolen a map to a large deposit of the resource.

He tells you to reclaim the map from their leader, Ancarion.

❖ Dealing With Ancarion

You can find the foreign operative on the north end of Solstheim, at Northshore Landing. He is waiting on a ship with another team of agents. One of his subordinates will stop you as you approach. If you ask to speak to the leader, you will get a chance to talk to Ancarion.

⋄ Persuading Ancarion

Ancarion will not be immediately hostile, though you must convince him if you want to resolve the situation without bloodshed. Persuading him that his attempt is a waste of time or intimidating him into leaving will let you take the map without needing to fight.

If you offer to make stalhrim weapons for him, he will give you the map to learn how to forge the weapons. Ancarion will remain on the island and purchase stalhrim weapons from you.

This final method is lucrative but will only be available if you have 75 Sneak. Failing to convince him of any option will have Ancarion turn hostile.

⋄ Defeating Ancarion

The Thalmor are a hard faction to love, and if you want to remove them from Nirn one agent at a time, you can attack them as soon as they are within reach. You will need to fight two Thalmor Guards and Ancarion himself.

Ancarion will attack you with magic and will likely summon a Storm Atronatch. He is no slouch and should be treated with caution. Make sure to loot his body for the map.

❖ Completing A New Source Of Stalhrim

With the map in hand, return to Baldor. He has returned to Skaal Village and will teach you how to use stalhrim as a reward for your help. If you would like to create stalhrim equipment, you will then have to source some.

⋄ Finding Stalhrim

Baldor will show you the Stalhrim Source, an otherwise unmarked area that allows you to gather 30 chunks of stalhrim at a time. With it, you will be able to craft the equipment.

You can only extract stalhrim with an Ancient Nordic Pickaxe, which can be purchased from Baldor or earned by completing a short quest for Glover Mallory.

You will be able to find more Stalhrim deposits across Solstheim at the many barrows that dot the island. While not as bountiful as the Stalhrim Source, they will help produce the items en masse.

Stalhrim equipment can be created at a forge, under its own category. It is available as both light and heavy armor and can be converted into all kinds of weapons. They are made with stalhrim, quicksilver ingots, steel ingots, and leather strips.

You will need the Ebony Smithing perk to make stalhrim equipment.

They are unique in their affinity for Frost. Enchantments that help you resist or deal more Frost Damage will be 25 percent stronger on stalhrim items.

Lost Legacy

❖ *Starting Lost Legacy*

Lost Legacy will only start after you complete A New Source of Stalhrim. You will have established a reputation as a reliable adventurer throughout the quest, prompting a historian, Tharstan, to approach you at its conclusion.

Tharstan was one of the locals who the Wind Stone had enthralled before completing the quest Cleansing The Stones.

He promises to pay you for protection as he explores newly unearthed sections in Vahlok's Tomb. Regardless of your response, he will wait for you there.

Vahlok's Tomb is south of Skaal Village and is easily accessed.

❖ *Solving Vahlok's Tomb's First Puzzle*

The newly uncovered section is mercifully free of enemies. It is a large, open area with several locked-off hallways with a fire pit in the middle. Tharstan is undeterred by the lack of an obvious path and gets to work on translating a pedestal.

To solve the puzzle, drag one of the nearby draugr corpses onto the grate covering the fire pit. Return to the pedestal and pull the handle located on the stem. It will open the grate and drop the body in, creating a sacrifice.

You lift bodies by holding the interact key while seeing the prompt to loot the body. This feature is rarely used in the game, with a notable exception being the Dark Brotherhood quest Recipe for Disaster.

With the puzzle solved, the two hallways to the north and south will open, allowing you to proceed. You can tackle these in any order, but we will start with the southern hallway.

❖ *Solving The Lost Legacy Pillar Puzzle*

The southern hallway has a handful of draugr that will try to stop Tharstan and you from venturing deeper. You will have to fight through them to reach the puzzle within this hallway. The route is linear and easy to follow.

Tharstan will cower in a corner during fights, keeping him safe. There is no need to worry about him when in combat.

The two of you will be stopped by a glowing pillar in the middle of a room. Its three sides are glowing with red, green, and blue energy respectively. There is also a small pedestal across from each face.

You must use the correct weapon against each pedestal to solve the puzzle.

Blue Pedestal	Use magic against it.
Red Pedestal	Use a melee weapon against it.
Green Pedestal	Use a ranged weapon against it.

Once the three pedestals have been lit, the way forward will unlock. You will need to fight several more draugr in the next room, including a powerful one that holds half of the Amethyst Claw.

Defeat all of them and claim the word for the Battle Fury shout and loot half of the Claw. Retrace your steps to explore the north hallway.

❖ Solving The Puzzle Grid In Vahlok's Tomb

The next puzzle is quickly accessed. It is a simple grid of pressure plates. You will need to step on all the plates without stepping on any of them twice. There are many solutions, though the one in the image above will work.

It will open the way to another Word Wall and another draugr holding the other half of the key. Defeat it, claim the word, and loot the key before returning to the central chamber.

❖ Crossing The Chasms In Vahlok's Tomb

At Tharstan's suggestion, insert each half of the claw into waiting receptacles at the westward gate. It will open a locked gate that starts the next challenge. You will need to pull a handle on a pedestal to start forming a series of magical platforms over a water-filled chasm.

You will have to run quickly to cross the path before each platform fades out of existence. Do not get too comfortable after your first success; the challenges will grow in speed and complexity, forcing you to adapt quickly each time.

Most of the chasms have Corrupted Shades similar to those from The Break Of Dawn in the water. Try not to fall in.

A permanent, safer path will form each time you complete a challenge, allowing Tharstan to cross safely.

❖ Finishing Lost Legacy

The final room is locked behind a Dragon Claw Puzzle. It is complicated by the halves of the claw lacking the solution on the palm. You will need to wait for Tharstan to provide the hints, or you could input the correct combination, indicated below.

Outer Ring		Bird
Middle Ring		Wolf
Inner Ring		Dragon

You will enter the final room. A Dragon Priest named Vahlok the Jailer will awaken to stop you. It uses Flame spells to fend you off, though it is not a particularly challenging enemy. With it defeated, you can claim the final word for Battle Fury and claim your gold from Tharstan, ending the quest.

How To Find And Beat The Ebony Warrior

❖ Where Is The Ebony Warrior?

The Ebony Warrior will appear in any major city (Whiterun, Solitude, etc.) once

211

your character reaches level 80, one level shy of Skyrim's original level cap. Enter any major city and a large NPC in full Ebony gear will speak with you, demanding an ultimate final fight against you.

If you cannot find the Ebony Warrior, try looking at the following locations:

- Whiterun: right after you enter the city. He should be near the Warmaiden's store.

- Falkreath: just outside the city gates on the southern end of town.

- Solitude: near the Hall of the Dead.

- Markarth: he will be standing outside of The Warrens.

- Riften: right next to Balimund's forge at The Scorched Hammer.

- Windhelm: he should be standing near Candlehearth Hall.

Once the Ebony Warrior has been spoken to, a miscellaneous quest named "Ebony Warrior" will appear on your screen, leading you to a location named Last Vigil. It is here where you will face the Ebony Warrior in combat.

❖ How To Beat The Ebony Warrior

Before you face the Ebony Warrior in battle, it is important to know just how powerful he is. This NPC has a wide range of strong perks, a rather large health pool, and some rather strong enchantments on his gear.

⬩   *Ebony Warrior Stats*

Health	2,071
Primary Skills (Rank 100)	One-Handed Archery Block Heavy Armor Sneak Conjuration Alteration Destruction Restoration
Perks	Most perks tied to his primary skills

	Extra Damage 3 (deals 3x as much damage)
	Paralysis immunity
Gear Effects	50% fire, cold, and lightning resistance
	40% increased health regen
	25 HP leech per sword hit
Miscellaneous	Uses Close Wounds, Ironflesh, Unrelenting Force, and Disarm

If it wasn't apparent from his stats, the Ebony Warrior has a massive amount of damage resistance against elemental damage. While blocking, he takes 75% less elemental damage. This makes defeating him by using spells somewhat difficult. The Ebony Warrior will also use the Disarm Shout if you stay near him.

◆ *Strategy #1: Stealth*

This boss is extremely susceptible to stealth. The Last Vigil is a rather spacious arena, allowing you to reposition and take him down with arrows from a distance. Crouch, hit him with an arrow, retreat until the Ebony Warrior's attention resets, then repeat.

If you want a faster and more humorous way of killing the Ebony Warrior, steal his items. If you have the Perfect Touch perk in the Pickpocket tree, you can take the Ebony Warrior's armor and weapons before the fight even starts. This will remove virtually all of his damage resistance, making this fight much easier.

◆ *Strategy #2: Shouting*

If your gear isn't strong enough to fight the Ebony Warrior, the Unrelenting Force Shout works well against him. A cliffside exists in the arena that you can fling the Ebony Warrior off of, dealing massive ragdoll damage to him. Make sure you damage him before he gets back up or the fall damage will be negated. Repeat this as needed to kill him.

◆ *Strategy #3: Crafting*

For a more direct approach, upgrade your weapons and armor as much as possible through Smithing and Enchanting. Get your crafting professions to rank 100, grab any perks that enhance the strength of items you make, then reinforce your weapons and armor to Legendary status. From there, apply two enchantments to each of your items.

Tip: Use the Marked for Death Shout on the Ebony Warrior. This will slowly strip all of his armor, increasing the damage of your weapons.

Absorb enchantments work best here since he is highly resistant to elemental damage. Chaos damage is also a good pick for your weapons. Armor enchantments will depend heavily on your build. With a hard-hitting weapon and high-ranking armor, you should be able to tank the Ebony Warrior's onslaught of blows while dealing a good bit of damage to him.

Strategy #4: Jarrin Root Poison

When you get near the end of the Dark Brotherhood quest chain, you'll be given an Alchemy ingredient called Jarrin Root. This item deals 200 damage to you when consumed, although its real strength is being able to combine with other poisonous ingredients to create the strongest poison in Skyrim. When combined with other ingredients with the "Damage Health" effect, you can craft a poison that deals 6,000+ damage instantly.

If you still have this ingredient, craft the strongest poison you can. Apply it to your choice of weapon, the hit the Ebony Warrior with it. If your Alchemy skill is high enough, this poison should instantly kill the Ebony Warrior.

Strategy #5: Magic

Using any close-range weapons against the Ebony Warrior is extremely dangerous. He can disarm you, use Unrelenting Force, and can even reflect damage towards you. Fortunately, he can't reflect spells.

Destruction and Conjuration spells are useful for fighting the Ebony Warrior. Conjuration spells allow you to summon one or two minions to draw his attention while you pelt at the Ebony Warrior from a distance with your Destruction spells. He is fairly resistant to magic, so this strategy will take slightly longer than the others on this list. Still, if you don't want to risk getting killed by reflected damage, this is a safe way of fighting the Ebony Warrior.

Made in the USA
Las Vegas, NV
30 January 2024

85115777R00118